CARBON SEQUESTRATION IN SOILS

Science, Monitoring, and Beyond

Proceedings of the
St. Michaels Workshop, December 1998

EDITED BY
Norman J. Rosenberg
R. César Izaurralde
Elizabeth L. Malone

BATTELLE PRESS
Columbus • Richland

Library of Congress Cataloging-in-Publication Data

Carbon sequestration in soils: science, monitoring, and beyond :
conference proceedings : St. Michaels Workshop, December 1998 /
edited by Norman J. Rosenberg, R. César Izaurralde, Elizabeth L. Malone.
p. cm.
Includes bibliographical references.
ISBN 1-57477-084-5 (alk. paper)
1. Soils—Carbon content Congresses. 2. Carbon sequestration Congresses.
I. Rosenberg, Norman J., 1930- . II. Izaurralde, R. César, 1948- .
III. Malone, Elizabeth L., 1947- .
S592.6.C35C37 1999
631.4'1–dc21 99-35376
CIP

Printed in the United States of America

Battelle Press
505 King Avenue
Columbus, Ohio 43201-2693
614-424-6393; 1-800-451-3543
Fax: 614-424-3819
Homepage: http://www. battelle.org/bookstore
e-mail: press@battelle.org

Contents

Executive Summary

Carbon (in the form of carbon dioxide) is accumulating in the atmosphere at a rate of about 3.5 billion metric tonnes per annum as the result of fossil fuel combustion, tropical deforestation and other land use change. The accumulation of carbon dioxide and other greenhouse gases in the atmosphere is expected to cause observable climatic changes in the coming century. The Kyoto Protocol to the United Nations Framework Convention on Climate Change, if ratified, will require developed nations to reduce their net carbon emissions in the period 2008-2012 to less than was emitted in 1990. Much effort is being devoted to developing strategies for reducing the emissions of the several greenhouse gases and in the case of carbon dioxide removing it from the atmosphere.

With regard to the option of removing carbon from the atmosphere, the Intergovernmental Panel on Climate Change (IPCC) has calculated that over the next 50 to 100 years, agricultural lands alone have the potential to remove anywhere from 40 to 80 billion metric tons. A workshop to explore issues central to achieving this degree of sequestration was organized by the Pacific Northwest National Laboratory, the Oak Ridge National Laboratory and the Council for Agricultural Science and Technology and was held at St. Michaels, MD, in December of 1998. The workshop was attended by nearly 100 Canadian and U.S. scientists, practitioners and policy-makers representing agricultural commodity groups and industries, Congress, government agencies, national laboratories, universities and the World Bank. Support for the workshop was provided by the U. S. Environmental Protection Agency, the U.S. Department of Agriculture, the U.S. Department of Energy, the Monsanto Company, and the National Aeronautics and Space Administration.

The St. Michaels workshop found that reductions in atmospheric carbon content can be achieved by large-scale application of tried-and-true land management practices such as reduced tillage; increased use of rotational crops such as alfalfa, clover and soybeans; and by an efficient return of animal wastes to the soil. Forests and grasslands afford additional capacity for carbon sequestration when established on former croplands. Programs to further soil carbon sequestration will provide ancillary benefits, including improvements in soil fertility, water holding capacity, and tilth; and reductions in wind and water erosion. However, although practical and economically viable farming methods that increase carbon storage in soil already exist, research is needed to develop new methods that increase the amount stored and the length of time carbon remains in the soil. Promising research leads involving applications of molecular science, colloidal chemistry, bioengineering and traditional plant breeding are described in Chapter 1 and the associated commentaries and working group report.

There is opposition to using soil carbon sequestration in the Kyoto Protocol calculations. One cause of the opposition is the perception that it will be difficult, if not impossible, to verify claims from around the world that carbon is actually being sequestered in soils. The workshop concluded that it *is* currently possible to monitor changes in soil carbon content, but current methods are expensive and not well suited to global monitoring. Technology, it was concluded, can provide new and widely applicable methods at a reasonable cost. These methods can be based on applications of remote sensing, direct non-destructive sampling, carbon flux monitoring in the field and the use of simulation models validated against observations. Details on these opportunities are given in Chapter 2 and its associated commentaries and working group report.

One special opportunity to link the objectives of two United Nations Conventions—the Framework Convention on Climate Change and the Convention to Combat Desertification—was discussed in considerable detail. The workshop concluded that there are vast areas of degraded and desertified lands throughout the world, many in developing countries, where improvements in rangeland management, dryland farming and irrigation can add carbon to the soil. Soil carbon sequestration can provide the impetus for changes in land management practices that will begin the essential process of stabilizing the soil against further erosion and degradation with concomitant improvements in fertility and productivity. Specific management practices, useful species and other appropriate steps in combating desertification are described in Chapter 3 and its associated commentaries and working group report.

Uncertainty about the costs, benefits and risks of new practices could impede the adoption of new technologies to increase carbon sequestration. The workshop concluded that financial incentives could increase the adoption of such practices and potentially provide an addition to farmer income. Government payments, tax credits, and/or emissions trading within the private sector are mechanisms that could be employed to overcome farmer reluctance. Impediment and adoption issues in need of clarification and research are discussed in Chapter 4 and its associated commentaries.

A number of recommendations emerging from drafts of the workshop papers and working group reports presented here have already affected agency planning with regard to soil carbon research. First among these recommendations is that targeted research in the basic science of soils and plants will yield understanding of how to enhance carbon sequestration in soils and reap the ancillary benefits of improved soil productivity, perhaps especially in desertified lands. Second, improvements in current monitoring and verification methods could include novel sampling technologies, computer modeling, eddy flux approaches, scaling up of data, and other methods such as remote sensing techniques. Third, the joint global goals of sequestering carbon and combating desertification could be facilitated through the use of mechanisms such as carbon trading, the Clean Development Mechanism, and an international carbon fund, enabling large-scale international projects and establishment of baseline data. Multiple environmental and productivity benefits could result from congruent efforts in basic science, monitoring and verification methods, and coordinated policy programs.

Introduction to Proceedings

Norman J. Rosenberg, R. César Izaurralde, Elizabeth L. Malone

Carbon in the form of carbon dioxide (CO_2) is currently accumulating in the atmosphere at the rate of about 3.4 Pg/yr (1 Pg=1 billion tonnes) as the result of fossil fuel combustion and land use change. The Intergovernmental Panel on Climate Change (IPCC) in its Second Assessment Report estimated that it may be possible, over the course of the next 50 to 100 years, to remove between 40 and 80 Pg of that carbon by sequestering it in cropland soils (Cole et al., 1996). Thus soil carbon sequestration on agricultural lands alone might offset the effects of fossil fuel emissions and land use change for one or two decades or even longer. Additional carbon sequestration is possible in managed forest and grassland soils. Article 3.3 of the Kyoto Protocol recognizes land use change and forestry (limited to afforestation, reforestation and deforestation since 1990) as valid source/sink activities that parties could use to meet greenhouse gas reduction commitments when reported in a verifiable and transparent manner. Yet managing soils for carbon sequestration was not included as a valid activity because of perceived uncertainties regarding verification of changes in carbon stocks.

The IPCC finding with regard to soil carbon sequestration has engendered excitement in many circles—including the Pacific Northwest National Laboratory's (PNNL's) Global Climate Change Group, which is working to develop a Global Energy Technology Strategy Program. This group undertook a preliminary analysis of the potential importance of soil carbon sequestration (croplands only and at the scale estimated above) as a component of a global strategy to constrain the atmospheric concentration of CO_2 within acceptable limits. Results of this analysis, made with the integrated assessment model MiniCAM 98.3 (Edmonds et al., 1996a,b), are shown in Figure 1. Calculations for the figure assume that from 2000 to 2100 soils sequester carbon at global annual rates ranging from 0.4 to 0.8 Pg/yr, with rates twice as great in the initial years and half as great in the later years. It is further assumed that the full potential of soil carbon sequestration is realized without any additional net cost (not unreasonable since the addition of organic substances to soil improves agricultural productivity).

Figure 1 shows that, overall, soil carbon sequestration can make a significant contribution to controlling the rise of CO_2 concentration in the atmosphere and, thus, to mitigating greenhouse warming. But, most strikingly, this analysis shows that, if the full potential of soil carbon sequestration can be realized in croplands, the time during which the global energy system can adjust to constraints that keep atmospheric carbon dioxide levels within acceptable bounds would be lengthened up to 35 years. The minimum present discounted cost of stabilizing the atmos-

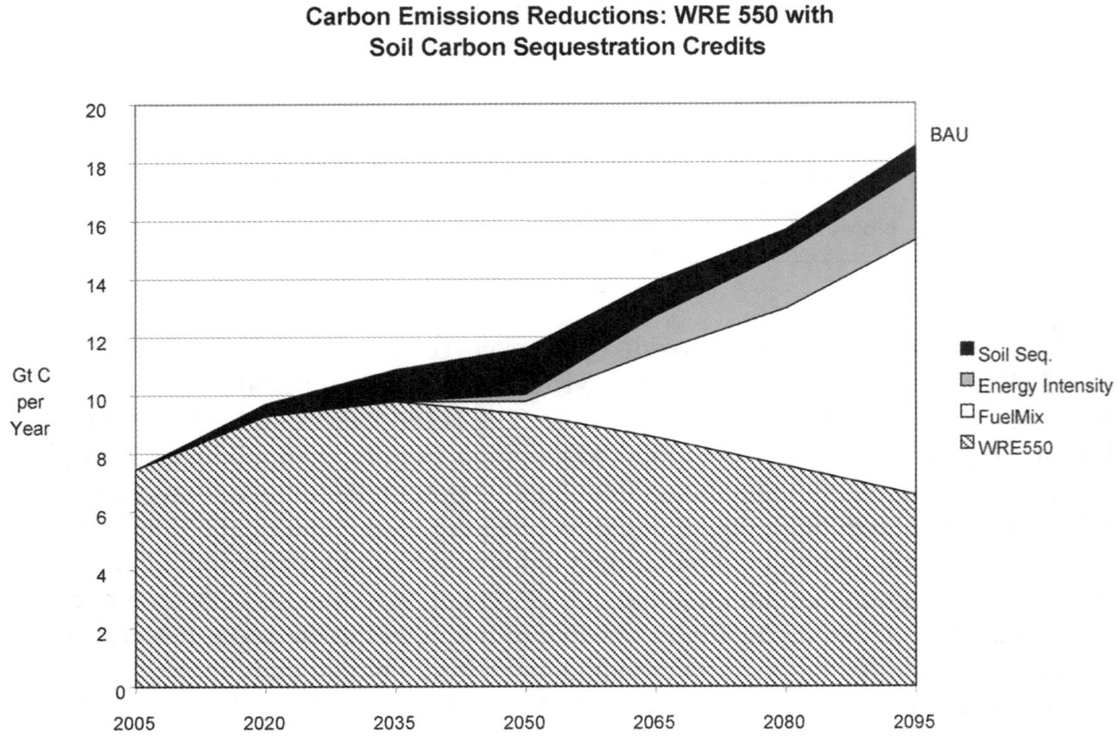

Figure 1. Global Carbon Emissions Reductions: WRE 550 (Wigley et al., 1996, 550 ppmv atmospheric CO_2 concentration). This figure shows a hypothetical path to carbon emissions reductions from MiniCAM's business as usual (BAU) emissions pathway to the WRE 550 concentration pathway, under a scenario in which credit for soil carbon sequestration is allowed. Soil sequestration of carbon alone achieves the necessary net carbon emissions reduction in the early part of the century. From the middle of the century on, further emissions reductions must come from changes in the energy system (such as fuel switching and the reduction of total energy consumption).

phere would be reduced in such a case by about 45% because (a) the degree of change required of the global energy system is reduced and (b) the changes are delayed until advanced energy technologies are ready for deployment. The power of discounting is also an important factor. Results of the MiniCAM cost calculations are illustrated in Figure 2. Had the effects of soil carbon sequestration in forests and rangelands been included, the MiniCAM analysis would show longer deferrals and still lower costs.

How realistic are the estimates of the potential for soil carbon sequestration on which the economic modeling is based? The estimates for cropland assume the restitution of up to two-thirds of the soil carbon released since the mid-nineteenth century by the conversion of grasslands, wetlands and forests to agriculture. The experimental record confirms that carbon can actually be returned to soils in such quantities: carbon has been accumulating at rates exceeding 1 tonne/ha/yr in former U.S. crop lands planted to perennial grasses under the Conservation Reserve Program (CRP) (Gebhart et al., 1994). Soil carbon increases ranging from 1.3 to 2.5 tonne/ha/yr have been estimated in experiments on formerly cultivated land planted to switchgrass, a biomass crop (preliminary data, Oak Ridge National

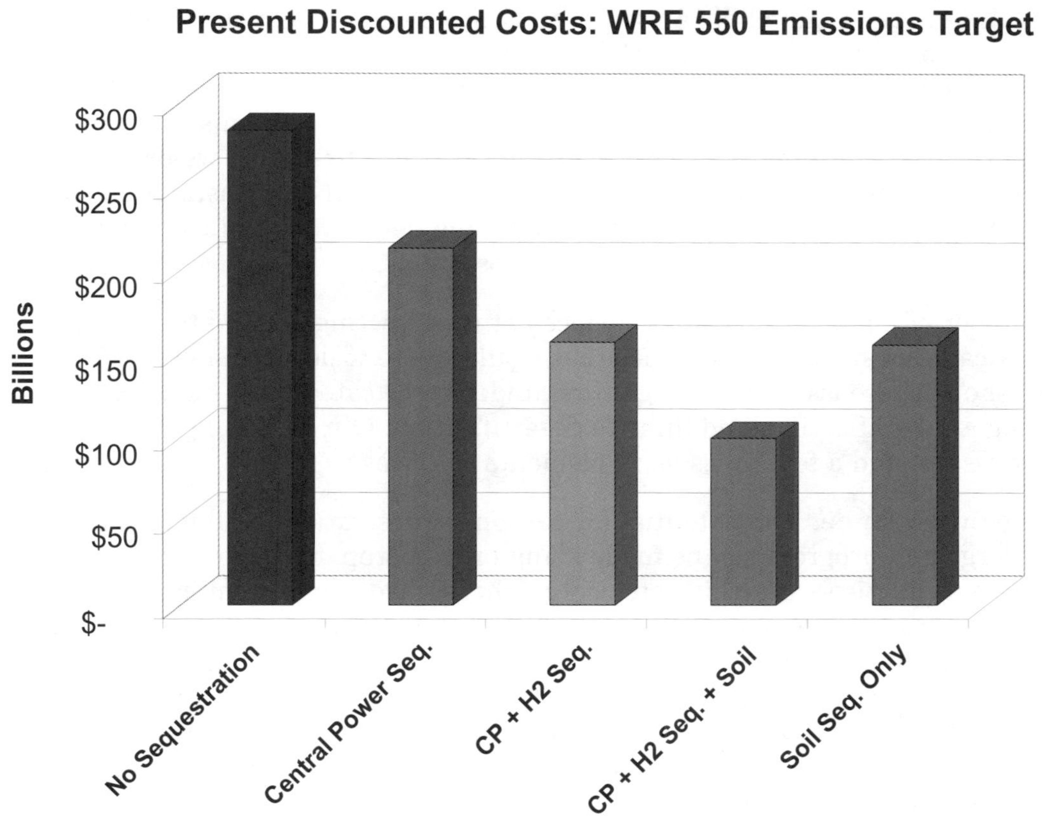

Figure 2. Present Discounted Costs: WRE 550 Emissions Target. This figure compares the present discounted costs (from now to 2100 at 5%/yr) of meeting the WRE 550 carbon emissions constraint under various assumptions about the possibility of carbon sequestration, as determined from a hypothetical scenario from the MiniCAM model. The cases shown from left to right are: no sequestration considered; sequestration of CO_2 captured at sites of central power production (CP); CP plus hydrogen fuel production (H2); CP + H2 plus soil carbon sequestration (soil) and soil sequestration only. Soil sequestration alone could reduce the discounted costs by about 45%.

Laboratory). Further, a substantial number of experiments over the last 20 years or so with minimum tillage and no-till management of farm fields have demonstrated that these practices lead to increases in soil carbon content (Lal et al., 1998).

Managed forests, wetlands and rangelands provide further opportunities for significant carbon storage. For example, when agriculture is converted (or allowed to revert) to forest vegetation in systems with very little management to improve growth, soil carbon may accumulate at rates ranging from near zero to 7 tonne/ha/yr. In temperate regions gains range from 0.2 to 0.6 and average about 0.3 tonne/ha/yr (Jenkinson, 1971). Gains in tropical and subtropical forests are greater, ranging from 1.0 to 7.4, with an average of 2.0 tonnes of carbon/ha/yr (Ramakrishnan and Toky, 1981).

With these facts and speculations before us PNNL's Technology Strategy Program proceeded to enlist the interest and aid of the Oak Ridge National Laboratory

(ORNL) and the Council for Agricultural Science and Technology (CAST) in a process to assess whether soil carbon sequestration can indeed be accomplished at the scale described above and, if so, what scientific and socioeconomic research will be needed to facilitate implementation of these practices on a large enough scale to make a difference. On April 23-24, 1998, as a first step, we brought an authoritative group of scientists (see Appendix I) to PNNL's Washington, DC, office to plan a major workshop on research and development of carbon sequestration technologies in managed terrestrial ecosystems.

A number of important findings surfaced at this meeting and led to identification of critical issues and opportunities that would warrant deeper analysis at the later workshop. These issues were organized under three categories—new science, monitoring and verification, and links to desertification. Our reasoning with regard to these issues, and a fourth issue, implementation, was as follows.

- **New Science:** The potential for carbon sequestration in all managed soils is large and progress can be made using proven crop, range and forest management practices. But this potential might be made even greater if ways can be found to restore more than two-thirds of the carbon that has been lost from conversion to agriculture and perhaps even to exceed original carbon contents in some soils and regions. This would involve a search for ways to effect greater, more rapid and longer lasting sequestration. New lines of research that promise to improve understanding of soil carbon dynamics and lead to improved sequestration methods should be explored.

- **Monitoring and Verification:** International agreements allowing soil sequestration to figure into a nation's carbon balance will require agreed-upon means of verification. Current methods are costly, time-consuming, and unable to distinguish year-to-year changes in soil carbon. Improved methods for monitoring changes in soil organic carbon might involve spatial integration based on process modeling and geographical information systems, application of high-resolution remote sensing, and continuous direct measurements of CO_2 exchange between the atmosphere and terrestrial ecosystems and/or the development of "carbon-probes."

- **The Soil Carbon Sequestration/ Desertification Linkage:** The potential for carbon sequestration is probably greatest on desertified and degraded lands. There are an estimated 2 billion hectares of such lands worldwide, 75% of them in the tropics with degradation most severe in the dry tropics. The linkages between loss of soil carbon and the desertification process are not fully understood. Nor are the mechanisms whereby restoration of carbon in soils can help to halt or even reverse desertification. The potential for increased sequestration on degraded lands through erosion control, agricultural intensification, forest establishment in dry regions, and biomass cultivation should be explored. Soil carbon sequestration could be a means of linking objectives of both the Climate Change and Desertification Conventions.

- **Implementation Issues:** Many soil carbon conserving practices are currently in use for such reasons as erosion control, increased productivity and/or reduced production costs. None are practiced thus far for the purpose of reducing the threat of global warming. Even for non-climate control applications there are impediments to more widespread adoption of carbon conserving practices. These include costs of initial investment in equipment, fear/reluctance to adopt new technologies, and increased risks during the adoption transition phase. Financial incentives such as government payments, tax credits, and/or emissions trading within the private sector might be used to increase the rate of adoption of carbon sequestering practices and potentially provide a significant addition to farmer income. So it was decided that at the workshop the key topics of New Science, Monitoring and Verification, and Links to Desertification (and others) would be examined, as well, from the point of view of Impediments to Implementation.

With this guidance before us a steering committee was appointed to organize the major Workshop. Committee members were David Shriner[1] and Gregg Marland of ORNL, Rattan Lal of Ohio State University, Richard Stuckey of CAST and Cesar Izaurralde and Norman Rosenberg of PNNL. First we decided that the workshop would be called "Carbon Sequestration in Soils: Science, Monitoring and Beyond." The next step was to enlist a number of leading scientists to identify the issues associated with each of the key topics in white papers to be distributed prior to the workshop. Then we recruited an interdisciplinary group of commentators and charged them with pointing out the strengths, weaknesses and omissions, if any, of each paper. The workshop agenda (Appendix II) was built around presentation of the papers and commentaries

While the papers were being written, the Steering Committee assembled a list of persons to be invited to the workshop. Although originally planned for 60 to 80, the workshop generated considerable interest; ultimately, about 100 people participated (Appendix III). Workshop participants represented a wide variety of backgrounds: natural and social scientists, agency program managers, policy-makers, farmers. They came from academia, national laboratories, the executive and legislative branches of government, nongovernmental organizations, business and industry. Most, if not all, of the participants agreed that the mix of viewpoints and interests (vested and otherwise) represented at the workshop made for stimulating argument and debate.

These proceedings contain the revised white papers. The commentaries on these papers are also included but have been edited slightly so as to highlight their key points. Although some or most of the commentators' suggestions were adopted by the white paper authors in their revisions, we leave them intact so that the reader can gain an understanding of the discussions and dynamics of the process used at the workshop. The reader may also judge how well the authors have dealt with commentators' criticism and suggestions.

All participants in the workshop had ample opportunity to discuss the issues raised in the papers and commentaries and to raise other germane issues. In addition, all participants at the workshop were assigned to breakout groups formed to

[1]Since transferred to the North Central Forest Experiment Station, St. Paul, MN.

study the new science, verification and desertification linkage questions. These groups reported their findings and recommendations to the plenary at various times. The working group reports are also included in these Proceedings.

The St. Michaels workshop and the publication of its proceedings were made possible by the support of four U.S. government agencies—the Department of Energy's (DOE's) Office of Basic Energy Research, the U.S. Environmental Protection Agency's (EPA's) Office of Policy, the National Aeronautics and Space Administration's (NASA's) Office of Earth Science, the Department of Agriculture's (USDA's) Office of Global Change and by the Monsanto Company. For their support and guidance and for their active participation in the workshop, we wish to thank the following persons in these agencies: Jerry Elwood and Roger Dahlman of DOE, Peter Kuch and Andy Manale of EPA, Charles Hutchinson of NASA, Margot Anderson of USDA and Bruno Alesii of Monsanto. Suzette Hampton and Laura Green of PNNL's Washington, DC, office handled all the logistical work involved in setting up and running the workshop. Participants were unanimous in praising their outstanding work.

REFERENCES

Cole, V., C. Cerri, K. Minami, A. Mosier, N. Rosenberg and D. Sauerbeck. 1996. Agricultural options for mitigation of greenhouse gas emissions. Chapter 23 in R. T. Watson, M. C. Zinowera and R. H. Moss (eds), *Climate Change 1995. Impacts, Adaptations and Mitigation of Climate Change*. Report of IPCC Working Group II, Cambridge University Press: 745-771.

Edmonds, J., M. Wise, H. Pitcher, R. Richels, T. Wigley, and C. MacCracken. 1996a. An integrated assessment of climate change and the accelerated introduction of advanced energy technologies: an application of MiniCAM 1.0. *Mitigation and Adaptation Strategies for Global Change* 1(4):311-339.

Edmonds, J., M. Wise, R. Sands, R. Brown, and H. Kheshgi. 1996b. *Agriculture, Land-Use, and Commercial Biomass Energy: A Preliminary Integrated Analysis of the Potential Role of Biomass Energy for Reducing Future Greenhouse Related Emissions*. PNNL-11155. Pacific Northwest National Laboratory, Washington, DC.

Gebhart, D.L., H.B. Johnson, H.S. Mayeux and H.W. Pauley. 1994. The CRP increases soil organic carbon. *Soil and Water Conservation* 49:488-492.

Jenkinson, D.S. 1971. The accumulation of organic matter in soil left uncultivated. Rep. Rothamsted Exp. Stn. for 1970, Part 2: 113-117.

Lal, R., J. Kimble, R. Follett and C.V. Cole (eds.). 1998. *The Potential of U.S. Cropland to Sequester Carbon and Mitigate the Greenhouse Effect*. Ann Arbor Press, Ann Arbor, MI.

Ramakrishnan, P.S. and O.P. Toky. 1981. Soil nutrient status of hill-agroecosystems and recovery after slash and burn agriculture (Jhum) in north-eastern India. *Plant and Soil* 60:41-64.

Wigley, T.M.L., R. Richels, and J.A. Edmonds. 1996. Economic and environmental choices in the stabilization of atmospheric CO_2 concentrations. *Nature* 379:240-243.

CARBON SEQUESTRATION IN SOILS

Science, Monitoring, and Beyond

1

Science Needs and New Technology for Soil Carbon Sequestration

F. Blaine Metting, *Pacific Northwest National Laboratory*
Jeffrey L. Smith, *U.S. Department of Agriculture*
Jeffrey S. Amthor, *Oak Ridge National Laboratory*

Beginning in the nineteenth century, the clearing and plowing of temperate forests and grasslands, the increased use of fossil fuels, and the ongoing deforestation of the tropics have led to a dramatic elevation in the atmospheric content of CO_2 and other so-called greenhouse gases. This is postulated by many scientists to be responsible for, or at least a contributing factor to a projected global warming trend of 2° to 5° Celsius over the next century (Houghton et al., 1996). If it is assumed that global warming of this magnitude is not desirable for social, environmental, economic or political reasons, then it is necessary to simultaneously reduce additions of CO_2 to the atmosphere and enhance its removal through different sequestration mechanisms. These actions include enhanced CO_2 uptake by, and control of emissions from terrestrial ecosystems. Two basic processes are relevant to carbon (C) sequestration: (1) CO_2 input through photosynthesis, and (2) longevity of assimilated C. Both are affected by climate and vegetation and there is some flexibility to modify these through human activity.

Enhanced terrestrial uptake over the next 50 to 100 years has been suggested as a way to at least reclaim the 150 or more Pg[1] carbon (C) lost to the atmosphere from vegetation and soil since 1850 as a consequence of land use change (Lal et al., 1998a; Houghton, 1995), thus effectively "buying time" for the development and implementation of new longer term technical solutions, including biotechnology and other approaches presented later in this chapter. However, the ultimate potential for terrestrial sequestration is not known because of a lack of fundamental understanding of (1) the biogeochemical mechanisms responsible for C fluxes and storage potential on the molecular, landscape, regional, and global scales, and (2) the complex genetic and physiological processes controlling key biological and ecological phenomena. Specifically, the structure and dynamics of the belowground *soil* component, which accounts for two-thirds of global terrestrial organic C

[1]One Pg (Petagram) = 10^{15} g = one billion metric tonnes = 10^{9} Mg = one Gt (Gigatonne).

stocks, is poorly understood. The purpose of this chapter is to consider innovative technology for enhanced sequestration and address the scientific issues that must be resolved to better understand soil C sequestration potential and apply appropriate management approaches.

BACKGROUND

Current Understanding of Soil Carbon Sequestration

The atmosphere in 1994 held about 750 Pg C as CO_2, corresponding to an average concentration of 358 parts per million by volume (ppmv). During the past decade atmospheric CO_2 concentration has been increasing at about 1.6 ppmv per year with the major anthropogenic sources being fossil fuel consumption and intensive deforestation. More than half of the ~8 Pg C added to the atmosphere each year by human activities (ca. 6.3 from fossil fuel burning and ca. 1.7 from tropical land use change) is taken up by and stored in terrestrial ecosystems and the ocean in both biological and chemical processes. For a number of years the annual disposition of 1.8 Pg of anthropogenic CO_2 emissions has been the subject of spirited scientific debate. This 1.8 Pg was termed the "missing sink" because its fate is not well understood (Gifford, 1994; Houghton et al., 1998). It is now thought that regrowth of forests in northern latitudes, specifically eastern North America, and a global CO_2 fertilization effect[2] probably account for much or most of this (Dixon, et al., 1994; Harrison and Broecker, 1993; Fan et al., 1998). What remains poorly understood are (1) the underlying biogeochemical processes that control C sequestration in the soil, and how these react to changing climate and land use, (2) the potential for soil as a C sink, and (3) whether there are effective ways, including advanced technology, for realizing or enhancing this potential in the coming decades.

On a global scale, soils are important both as a source and sink of C. Therefore, it is important to acknowledge both enhanced sequestration *and* reduced emissions when considering soil C management. Global terrestrial photosynthetic C fixation is estimated to be about 120 Pg C/yr, at least half of which ends up in belowground root and soil components (Oades, 1988). The total amount of stored C in terrestrial ecosystems is about 2000 ± 500 Pg.[3] With a global stock of about 1500 Pg (~1393 Pg, Post et al., 1982; ~1576 Pg, Kimble et al., 1990) of soil organic C (SOC), soil holds 2.1 times as much C as the atmosphere and 2.7 times more than standing vegetation. Estimates of soil inorganic C (SIC) are less accurate although global stocks are estimated to be about 12% larger than SOC (Schlesinger, 1997). Regardless of whether a given soil is a C source or sink, the current and potential C storage capacity and annual C fluxes vary significantly among regions and ecosystems. Of the total land surface of about 15 × 10^9 ha, approximately 3-4 × 10^9 ha (22%) is potentially arable with irrigation. An estimated 40% of this

[2] Research has shown that net primary production (NPP) increases asymtotically with increasing atmospheric CO_2 concentration—the so-called CO_2 fertilization effect, as discussed later in this chapter.

[3] Soil to 1m depth plus vegetation, exclusive of peat, bogs and surface litter, which probably elevates the total to over 2500 Pg.

(1.46×10^9 ha) was in cultivation in 1990. At the same time, the influence of erosive forces on global C movement and storage is considerable. Farming, deforestation, and over-grazing results in erosion at an annual rate of about 26×10^9 tons of soil. The area of degraded land is approximately 2×10^9 ha (Lal, 1990; Lal et al., this volume).

The role of soils and soil organic matter (SOM) in the global C cycle has been the focus of extensive research for many years. In contrast, less attention has focused on SIC, which exists largely as carbonates in arid and semi-arid regions in soils derived from calciferous parent materials and in lime deposits (Schlesinger, 1985, 1992). Globally, the soils of arid and semi-arid regions hold 800 Pg of calcium carbonate ($CaCO_3$) in calcic and petrocalic horizons known as caliche, which accumulates at rates of 0.12 to 0.60 g C m^2/yr (Schlesinger, 1997). Applying this rate to all deserts (18.2×10^{12} m^2) we can estimate an upper annual C fixation limit of 0.011 Pg for this sink. The rate of $CaCO_3$ deposition in deserts is sensitive to CO_2 concentrations in the soil profile and to soil water content, both of which are affected by plant metabolism and by the existence of a calcium source (Schlesinger et al., 1987). Processes that influence or control SIC formation and movement, such as soil respiration, are poorly understood, particularly for non-arid regions.

The formation, distribution, and turnover of SOM are largely controlled by climate (i.e., temperature and moisture) and modified by vegetation, parent materials, and human activity. SOC is allocated over time to different "pools" as a consequence of root growth and subsequent decomposition, litter fall and decomposition, microbial degradation and synthesis, mixing by soil fauna, and moisture and temperature cycles. The pools are variously defined on the basis of relative recalcitrance which, in turn, governs residence and turnover times. One scheme (Eswaran et al., 1995) defines four pools based on C dynamics:

- **Active or labile pool** of readily oxidized compounds. Factors controlling the formation and dynamics of this pool include plant residue inputs and climate. Agronomic management practices affect the size of this pool.
- **Slowly oxidized pool** associated with soil macroaggregates. Factors controlling this pool are soil physical properties, including mineralogy and aggregation. Agronomic practices also affect the size of this pool.
- **Very slowly oxidized pool** associated with soil microaggregates. The controlling factor is water stability of the aggregates. Agronomic practices have little effect on this pool.
- **Passive (recalcitrant) pool.** Clay mineralogy is the controlling factor for this pool. Agronomic practices probably do not affect this pool.

Carbon residence times range from days for the labile pool to decades and centuries for the very slowly oxidized and passive (recalcitrant) pools. Charcoal constitutes another large recalcitrant pool whose residence time is independent of soil mineralogy.

The soil active fraction is usually defined as the soil microbiota plus the labile pool of C. In general, rates of soil metabolism are C/energy limited and thus dependent on periodic inputs of surface litter and root products. The kinetics and overall metabolic functioning of the active fraction is highly variable and controlled by abiotic parameters of temperature and moisture and by substrate quality. The C/energy flux through this fraction is the driving force for the decomposition of litter input and detrital material. If the decomposition rate is equal to organic matter input, the system will maintain steady state. However, if decomposition rates exceed litter input, residual organic matter will decline and both C and nitrogen (N) will be lost from the system.

On a global basis the active fraction has been estimated to be 6-9 Pg C and contain 0.9 to 1.3 Pg N (Smith and Paul, 1990). This fraction typically constitutes 2 to 5% of the total C of surface soils and decreases exponentially with depth. It is very labile and in most ecosystems turns over (regenerates) rapidly. Depending upon the quality of substrate and accounting for yield and respiration, a turnover of 5 to 10 times a year would consume 30 to 90 Pg C just to maintain this fraction on a global basis. Thus, in the short- to mid-term of years to decades, it is extremely important for purposes of managing soil C sequestration to better understand dynamics of the active fraction.

It is evident based on this analysis that soils, through the active fraction, play an important role in global C storage, and our lack of understanding of processes that maintain these pools makes analysis of C sequestration and global change difficult and uncertain. The active fraction is heterogeneous and so can vary locally by an order of magnitude within the same ecosystem. This leads to large uncertainty in estimated global SOC stocks that are derived from unrelated studies and inventories that are unrelated and were not originally designed to represent large areas of the Earth. And yet, as a percentage of total soil C, the active fraction seems relatively consistent across ecosystems. This implies that mineral soils may have an inherently limited capacity to maintain the active fraction. With the exception of bogs and peat-based soils, it also implies that whatever pool of C is controlling the active fraction may be similar across soils and ecosystems. Mineral soils may also have a limited capacity to protect or sequester C; otherwise, total SOC would not range from 0.5% to 8%; it might range between 10 and 20% or some other percentages.

Soil texture and structure also influence sequestration and allocation of C. Texture refers to the size distribution of the primary mineral constituents—sand, silt, and clay. Studies with many soils have shown that SOC preferentially adsorbs to clays, which include the finest-sized particles down to colloidal dimensions. Clay-associated SOC is generally more recalcitrant. Structure refers to three-dimensional aggregate properties of soil. Structural stability is the ability of a soil to retain its three-dimensional arrangement of aggregates and pore space in the face of wetting and drying, freeze-thaw cycles, temperature fluctuations and mechanical disturbance by roots, soil fauna, or tillage practices (Kay, 1998). A number of factors influence structure, including climate, parent materials, and microbial activity (Lynch and Bragg, 1985). Conceptually, the soil fabric comprises a hierarchy of aggregation. Larger, macroaggregates are bound together by transient polysaccharides, roots, and fungal hyphae. Macroaggregates are, in turn, composed of smaller microaggregates that are stabilized by persistent organic materials (humic substances) and by clays (Tisdall and Oades, 1982). Organic matter associated with microaggregates is largely in the passive (recalcitrant) pool.

SOIL CARBON SEQUESTRATION POTENTIAL

Native Ecosystem Potential

Land use options for enhanced C sequestration at the landscape and regional scales include protection and selective management of native ecosystems,[4] and use of appropriate and advanced management practices in manipulated ecosystems, which is the topic of the next subsection. Of course, there is not always a clear distinction between these two approaches, with constructed wetlands, managed grazing of rangelands, and "multiple use" national forests being some examples.

As a baseline for judging C sequestration potential, an understanding of inherent ecosystem productivity is important. The ratio of net plant growth (gross production - respiration) to the amount of absorbed photosynthetically active radiation (PAR) is often called the light (or radiation) use efficiency.[5] In an analysis of the literature, Ruimy et al. (1994) derived values for the light use efficiency (ε) for a number of ecosystem types (Table 1-1). The values range widely, being especially high for C-4 crops[6] and low for some forest types. Establishing accurate values for ε is difficult, so that the estimates in Table 1-1 should be interpreted with caution. There is also considerable variation among individual studies of the same ecosystem type. Nonetheless, differences among ecosystem types are apparent, and these differences highlight the importance of above-ground vegetation in the C balance of the terrestrial biosphere and for understanding C sequestration potential. In particular, the light use efficiency of managed ecosystems seems to exceed that of unmanaged native ecosystems. Whereas light use efficiency need not be strongly related to long-term C sequestration, it does represent the input side of the C balance equation for a given light environment and canopy structure.

Differences in canopy photosynthetic capacities among ecosystem types contribute to the importance of land cover or land use change on (potential) ecosystem CO_2 capture and C sequestration. Instantaneous photosynthetic responses of plant canopies to light differ among ecosystem types. Based on a review and analysis of micrometeorological and canopy-enclosure measurement data, Ruimy et al. (1994) concluded that photosynthetic capacity (i.e., CO_2 uptake at high irradiation, $\sim$1800 μmol photons/(m^2 sec) is greatest in crops, followed by grasslands and forests (Table 1-2). The lower apparent photosynthetic capacity in C-4 grasslands compared to C-3 grasslands, in spite of the generally greater leaf-level photosynthetic capacity in C-4 species, may be attributable either to the limited data from

---

[4] This has been referred to as "enhancing the natural C cycle" (Benson et al., 1997).

[5] ε = g dry matter per megajoule absorbed photosynthetically—active radiation

[6] Plants relying exclusively on the classical Calvin cycle are termed C_3 plants because the first product of CO_2 fixation is a three-carbon compound called phosphoglyerate. In contrast, C_4 plants produce dicarboxylic acids (four-carbon compounds) as the initial products of CO_2-fixation (Sage and Monson, 1998).

Table 1-1. Light Use Efficiencies among Ecosystem Types (Ruimy et al., 1994)

Ecosystem type	Mean ε (g/MJ)*
Equatorial moist forest	0.62
Equatorial evergreen tree plantation	1.74
Mediterranean evergreen forest	0.37
Temperate deciduous forest	1.01
Temperate deciduous tree plantation	2.72
Temperate and subpolar coniferous forest	1.57
Temperate and subpolar coniferous tree plantations	1.69
Temperate grassland	1.26
C-3 crops	2.71
C-4 crops	3.51
All cultivated/managed vegetation	2.07

* With respect to total (i.e., above- plus below-ground) plant growth

Table 1-2. Instantaneous Photosynthetic Capacity (CO_2 uptake at PPFD = 1800 μmol m^2/s) of Plant Canopies in Different Ecosystems (Ruimy et al., 1994)

Ecosystem	Photosynthetic Capacity μmol CO_2/m^2/s
Broadleaf forests	21
Conifer forests	19
C-3 grasslands	31
C-4 grasslands	23
C-3 crops	27
C-4 crops	33
All forests	20
All grasslands	24
All crops	28

grasslands or differing degrees of stress experienced by the measured C-3 and C-4 grasslands.

Table 1-3 includes speculative estimates of the potential for sustained terrestrial (soil + vegetation) C sequestration in native and managed ecosystems developed by a group of experts engaged in a current exercise sponsored by the Department of Energy (1999). The relative potential for C gain in native ecosystems in the next few decades is probably greatest in tropical savannah, followed by tropical forests, wetlands, and unmanaged temperate grasslands and forests (approximately equivalent). With hypothesized global warming, the potential for significant loss of soil C is thought to be greatest in high latitude ecoregions, including peat lands and arctic and boreal tundra and taiga (Amthor and Huston, 1998).

Management of Croplands, Forests and Grasslands

It is generally thought that soils have a finite C carrying capacity, dictated ultimately by the interactive temperature and moisture components of climate. In theory, when ecosystems are supporting climax plant communities, the annual net primary productivity (NPP) is balanced by the decomposition of residues and SOM and the system is in C equilibrium. At this point the soil will be at its maximum native carrying capacity for C. However, for managed ecosystems, it is possible to increase the soil carrying capacity for C through species selection or by altering the microclimate via nutrient and water management and other means. Two principal approaches for increasing terrestrial C sequestration are converting marginal land to grasslands and forest, and increasing productivity on crop and forest land with residue management and application of technology. Because the technology is in place today, the full sequestration potential of managed systems could be achieved with modest economic incentives.

On a global basis it is difficult to estimate the potential for increasing C sequestration because baseline inventory data are inadequate; the difficulty is compounded by varying and uncertain use of terms such as "marginal" cropland, "grassland" and "degraded" soils. We present here an analysis of the potential for C sequestration on U.S. soils for which there are adequate data on land use and soil C inventories. Based on such information, conclusions regarding C sequestration potential of managed systems should be applicable wherever in the world local land use and economic conditions are known. Because the best options for increasing C are those for which good management practices already exist, forests and croplands can be usefully evaluated to address the consequences of changing land uses. Thus, the discussion centers on manipulation of forests, agricultural land and, to a lesser extent, grazed and ungrazed grasslands.

There are 47 Mha[7] of marginal cropland in the United States, 14 Mha of which have been converted to grassland under the Conservation Reserve Program (CRP). Thirty-three Mha remain available for conversion or restoration to high C sequestrating vegetation, with the only constraint being guaranteed profitability to the

[7]Mha = million hectares. One hectare ~ 2.47 acres.

Table 1-3. Sustained Terrestrial C Sequestration Potential. The primary C sequestration method is rated with High (H), Medium (M), and Low (L) levels of sustained management intensity required over the long term. Global potential C sequestration (CS) rates were estimated that might be sustained over a period of up to 50 years. From the DOE Soils & Vegetation C Sequestration Roadmapping Team (G.K. Jacobs et al.).[1]

Ecosystem	Primary Method to Increase CS	Potential CS (GtC/y)[2]
Agricultural lands	Management (H)	0.85 – 0.90[a]
Biomass crop lands	Manipulation (H)	0.5 – 0.8[b]
Grass lands	Management (M)	0.5[c]
Range lands	Management (M)	1.2[d]
Forests	Management (M)	1 – 2[e]
Wetlands	Restoration, creation and maintenance (M)	0.1 – 0.2[f]
Urban forest and grass lands	Creation and maintenance (M)	---[g]
Deserts & Degraded lands	Manipulation (H)	0.8 – 1.3[h]
Sediments and aquatic systems	Protection (L)	0.6 – 1.5[i]
Tundra and taiga	Protection (L)	0.1 – 0.3[j]
TOTAL		5.65 – 8.71

[1]U.S. Department of Energy, 1999. Chapter 4. Carbon Sequestration in Terrestrial Ecosystems.
[2]R&D allows improvements in carbon sequestration implementation. Management intensity includes fertilization, irrigation, pesticides, and heavy equipment usage. No reallocation of land use, except for 10-15% of agricultural land to biomass crop lands. Baseline values are from Amthor and Huston (1998).

[a]Soil carbon only; Recovery of an amount equivalent to what was lost from native soils prior to agricultural use; Implementation of best-available management (e.g., no-till, intensified production and residue inputs, intensified rotations with crop rotation, double cropping, greater use of perennials) and new technologies, such as precision farming.
[b]An average annual aboveground productivity level of 13.2 MgC/ha/y. Belowground C storage is 1.75 Mg/ha/y and assumed to be "permanent" and to not provide any negative feedback on further storage; Short rotation woody crop and perennial grass production is assumed to provide equivalent C storage benefits; The energetic costs of producing and harvesting switchgrass results in a biomass energy return ratio (energy in harvested biomass divided by production energy costs) of 12.3 and an energy gain of 343% for ethanol production. The C gain from substitution of ethanol for gasoline (2.48 MgC/ha/y) after subtracting carbon costs of production (0.60 MgC/ha/y) and adding an average below-ground sequestration rate of 1.75 MgC/ha/y provides annual C savings of (2.48 + 1.75 – 0.60) = 3.60 MgC/ha/y; Trees and grasses are assumed to be equally efficient at net C production and sequestration and that production of ethanol and electricity provide equivalent net benefits in terms of C savings; A conversion of 10% of current cropland to biomass crops for energy represents a realistic target, while under more favorable conditions a 15% conversion might be achievable on a world basis.
[c]Intensification of management with fertilization, controlled grazing, and species improvments; 25% increase in below-ground carbon stocks; Linear increases through 2050.
[d]Total increase of 27 GtC through 2050; Rehabilitation of degraded range land and fertilization by increasing CO_2.
[e]Watson et al. (1996) estimate 1-1.6 GtC/y (their Table 14) and include above- and below-ground vegetation, soil C, and litter. Their estimate does not include R&D to increase carbon sequestration. Trexler (1998) suggests a rate of 2 GtC/y may be plausible. With focused R&D, both these values may be exceeded.
[f]The wetlands estimate is from Armentano and Menges (1986). It is for temperate and boreal wetlands, and it primarily represents C lost to disturbance of wetlands by agriculture, forestry and peat harvesting. The range for the tundra was taken from Oechel et al. (1993) and represents the impact of regional warming on net C balance. Thus, the two estimates may overlap in global coverage, but the first represents direct anthropogenic disturbance that is presumably reversible, and the second represents an indirect impact that would be difficult to manage. Tropical wetlands are not included in the estimate.
[g]No estimate available.
[h]From Table 3-24 of Lal et al. (this volume). Soil C emphasis; erosion, desertification and global warming effects are controlled; includes restoration of lands, reclamation of salt-affected soils, agricultural intensification on non-degraded lands (~.015 GtC/y), and fossil fuel C offset of ~0.2 GtC/y; includes accretion of inorganic carbonates.
[i]Estimate from Stallard (1998).

landowner. Current CRP land is concentrated in the mid-U.S. from Texas to Minnesota.[8] While the principal CRP benefit is reduced soil erosion, ancillary benefits, such as enhanced wildlife habitat, improved air quality, and improved surface water quality make the estimated benefit of this program $3.4 to $11 billion annually (Young and Osborn, 1990). The C sequestration potential of CRP land has been estimated to range between 0.4 and 1.0 metric tons (MT) C/ha/yr, for a total annual sequestration potential of between 6 and 14×10^6 MT C (0.006-0.014 Pg).

It has been estimated that about 25 of the 47 Mha of degraded cropland and pastureland in the U.S. are suitable for growth of softwood tree species and 22 Mha for hardwoods (Parks, 1992). Some of this land is taken up in the CRP so that the following applies to the remaining 33 Mha. Approximately 10 Mha were used for an analysis of softwoods because economic data were available to indicate this conversion would maintain a 4% annual return on investment. Considered a low estimate, the analysis suggested an annual C storage increase similar to CRP land of 0.033 Pg C (~0.57 MT C/ha). As shown in Table 1-4, if all available land were converted with similar production rates (biological potential) the annual C sequestration would be 0.119 Pg C.

Table 1-4. Annual U.S. potential for C sequestration from managed forests, arable lands and pastures.

Strategy	Average C sequestration	
	Low estimate	High estimate
	------------ Pg C / year ------------	
Forestry[1]		
Converting marginal crop/pasture to forest	0.033	0.119
Increasing timber growth on timber land	0.138	0.190
Growing short-rotation woody crops for energy	0.091	0.180
Increasing tree numbers/canopy cover in urban areas	0.011	0.034
Planting trees in shelter belts	0.003	0.006
Subtotal	0.276	0.529
Arable land[2]		
Cropland conversion to CRP (excluding agroforestry)	0.006	0.014
Soil restoration (eroded land, mine land, salt affected soil)	0.011	0.025
Conservation tillage/residue management	0.035	0.107
Better cropping systems (fertilizer, cover crops, manure)	0.024	0.063
Subtotal	0.075	0.208
Total managed forests, arable land, pastures	0.351	0.737

[1] after Hair et al., 1996; Birdsey, 1992.
[2] Lal et al., 1998b.

[8]Thirteen states account for over 75% of the acres in CRP, a $2 billion conservation program.

The second overall approach for enhancing C sequestration is to increase the productivity of crop and forest land. In agriculture it has been shown that high yields can be maintained while simultaneously managing land for optimum C storage in the form of increased surface residue and SOM. In this respect the adoption of conservation tillage practices is a viable mechanism to increase C storage by (1) reducing erosion, (2) increasing soil aggregation, and (3) decreasing the loss of SOM to microbial oxidation that results from tillage (Lal et al., 1998b). However, the adoption of conservation tillage has been somewhat slow because yields can be depressed in the first few years. Research is finding ways to overcome these problems and, as of 1997, 37% of all U.S. acreage was under some form of conservation tillage. The C sequestration potential of the combined practices of no-till, mulch, and ridge tillage was estimated to be 14.1 × 10^6 MT (0.014 Pg) C/yr, with associated savings in fossil fuel equivalent to 1.6 × 10^6 MT C/yr. In addition, managing crop residues from these systems may sequester another 22.5 × 10^6 MT (0.023 Pg) C/yr (Cole et al., 1996; Lal et al., 1998b).

The U.S. Forest Service has estimated that 85 Mha of forest land have the biological potential to increase production through regeneration and stocking control. Vasievich and Alig (1996) showed, for an economic constraint of a 4% annual return on investment, that these timberlands could sequester 0.138 Pg C/yr under proper management. They also estimated that, if all timberlands in this analysis, regardless of economic constraints, were managed for C sequestration, the potential would increase to 0.19 Pg C/yr (Table 1-4).

Another strategy for sequestrating C is the conversion or reallocation of agricultural land to woody crops. The U.S. Department of Agriculture (USDA, 1990) projected that 52 Mha of agricultural land will be in excess of demand for food production from 2000 to 2030. Wright et al. (1992) estimated that between 14 Mha and 28 Mha of cropland are suitable for woody crops. With a current production level of 6.5 MT C/ha, the sequestration potential for this strategy is 0.09 to 0.18 Pg C/yr (Table 1-4).

Other sequestration potential for both forestry and agriculture is also included in Table 1-4. Lal et al. (1998b) estimated a total potential for agriculture between 0.075 and 0.208 Pg C/yr. The totals for forestry and agriculture range between 315 and 633 × 10^6 MT C/yr (mean of 0.47 Pg/yr). The lower and upper total values in Table 1-4 represent 22% to 44% of the total U.S. fossil C emissions of 1.44 Pg, as estimated by the U.S. Department of Energy, and slightly less when using the total reported by the U.S. Environmental Protection Agency (1.709 Pg).

Grasslands constitute approximately 26% of the total U.S. land area, equivalent to 240 Mha, in comparison to cropland (134 Mha) and forests (298 Mha). On a worldwide basis temperate grasslands occupy 900 Mha. Grasslands are some of the most productive systems in the world with annual NPP of 467 gC/m^2 compared to savanna (313 C g /m^2/yr), temperate forests (558 g C/m^2/yr) and tropical forests (832 g C/ m^2/yr) (Burke et al., 1997).

Native grasslands in North America are designated as “mixed grass,” “short grass,” and “tall grass” prairie. Coupland (1992) estimated net above-ground production for the mixed grass prairie to range from 250 to 600 g/m^2/yr, with below-ground production ranging from 329 to 1024 g/m^2/yr. In the short grass steppe, the above-ground NPP was estimated at 50 to 325 g/m^2/yr, and below-ground between 540 to 790 g/m^2/yr (Lauenroth and Milchunas, 1992). Kucera (1992) collated information on tall grass prairie and reported a range of 200 to 1000 g/m^2/yr, with below-ground production being at least equal to above-ground biomass production.

Worldwide, temperate grasslands have similar productivity ranges. To increase production, it is necessary to bring these lands under management with the option of choice being N fertilization. This could potentially increase NPP significantly. However, grasslands have an inherent capacity to emit N_2O, a strong greenhouse gas (Mummey et al., 1999), and this would likely increase with the use of N fertilizer. Thus, a comprehensive analysis that includes NO_x emissions and the C cost of fertilizer production and application is needed to evaluate the net C sequestration benefits of fertilization of grasslands.

Degraded soils also represent a large potential for C sequestration. Worldwide, there are approximately 1965 $\times$ 10^6 ha of degraded soils, 4% from physical degradation, 56% from water erosion, 28% from wind erosion, and 12% from chemical degradation. With proper management these soils have the combined potential to sequester between 0.81 and 1.03 Pg C/yr. The best options include reclamation of saline soils, erosion control, restoration of eroded lands, and biofuel production (Lal et al., this volume).

The Role of Models

Soil process models have been used to describe the transfer of material among different C pools (Smith and Paul, 1990). The early aim was to predict N transformation and release, an important plant growth parameter. These models matured into numerous general SOM transformation models (Paustian et al., 1997). Mechanistic process models can be useful for describing net C sequestration, NPP, and greenhouse gas production from soil. However, where the processes and limitations involved are poorly understood or the ecosystem or land area is not well defined, the utility of mechanistic models is lost and simple bookkeeping models may suffice.

More complex models have also been developed to predict plant biomass production; and the natural marriage of plant and soil sub-models led to development of systems-type models (Groot et al., 1991). A few SOM and system models use explicit representation of microbial communities to transform litter to SOM. However, the use of this explicit interaction is not widespread. Most models use implicit calculations for sub-model transfers of material, such as C and N. Irrespective of the equations used in the numerous processes and transfers, most soil models have similar features such as inputs of temperature and moisture as controlling factors, C and N stocks, soil texture, and partitioning of litter into quality components for organic matter production (Potter and Klooster, 1997). In addition, SOM models are governed by rate functions, decomposition efficiency, and pool transfers (Hyvnen et al., 1998). Post et al. (this volume) describe the characteristics of four processed-based soil organic matter models and the need to apply them at regional scales. We next describe regional to global aggregated models used to evaluate the potentials of soils and vegetation to sequester C.

It is important to realize that existing data limit the use of models for representation of temporal and spatial variability, stock and flux values among ecosystems, and factors governing mechanistic processes across ecosystems. As Dai and Fung (1993) stated, “It should be emphasized that our biospheric carbon fluxes were calculated from simple empirical models using data having errors. Thus the conclusions drawn from them, especially those related to net carbon fluxes, are only suggestive.”

A number of models have been used to evaluate soil and vegetation C sequestration potential on a regional-to-global basis and to estimate the effects of management practices and land use change on C storage. The range and scope of the models vary widely, but most now acknowledge the value of simultaneous consideration of C and N fluxes (Schimel et al., 1994). For example, using the Terrestrial Ecosystem Model (TEM), McGuire et al. (1992) predicted that with a 2° Celsius rise in temperature, NPP would decrease if the C and N cycles were uncoupled, but increase when the cycles were coupled.

There are many recent additions to the literature from which to compile an overview of the status of global C cycle models. These include Siegenthaler and Sarmiento (1993), Post et al. (this volume), Bouwman (1990), Wollast et al. (1993), Eddy and Oeschger (1995), Lal et al. (1995a, 1995b), and Walker and Steffen (1996). A few model descriptions follow to give perspective.

McGuire et al. (1992) used the TEM to evaluate the influence of C/N cycle interactions on the prediction of NPP for 17 non-wetland vegetation types in North America. This process model considers three pools and uses spatially referenced climatic, soil, and vegetation data: C and N in vegetation, C and N in soil, and inorganic soil N. There are eight fluxes in the model: (1) gross primary production, (2) plant respiration, (3) C in litter production, (4) soil respiration, (5) N input, (6) N uptake by vegetation, (7) N in litter production, and (8) N lost from the system. The authors modified the original model by letting the C/N ratio of the vegetation vary, thus producing more realistic feedbacks between C and N. Melillo et al. (1993), used TEM to estimate patterns of global NPP; the model output was correlated to some degree ($r = 0.69$) to observed NPP. They next used the model to evaluate the influence of elevated CO_2 on NPP and concluded that process-based models such as TEM were useful for both scientists and policy makers interested in the potential effects of climate change on the biosphere.

Potter et al. (1993) used the Carnegie-Ames-Stanford (CASA) Biosphere model to estimate NPP by latitude, litter storage and soil C turnover. On a monthly time step, CASA simulates seasonal plant C fixation, biomass and nutrient allocation, litter fall, soil N mineralization and microbial respiration. Major inputs to the model are solar radiation, temperature and precipitation, soil texture, and C and N pools. The authors concluded that while estimates of NPP were comparable to other models, CASA was not ideally suited for studying land use, atmospheric processes, or climate change. However, they were successful in estimating C and N in litter and soil pools for various global ecosystems. The analysis is potentially useful for modeling shifts in C and N pools when the lignin content and/or litter quality varies.

Hudson et al. (1994) developed a global carbon cycle model (GLOCO) using a highly aggregated representation of oceans and terrestrial ecosystems. This moderately complex process model simulates the chemical and biological processes in the ocean with cycling of inorganic and organic C, phosphate and alkalinity. The terrestrial sub-model divides the biosphere into seven biomes with dynamic C and N cycling. In addition, the model incorporates anthropogenic effects such as fossil fuel combustion and agricultural development. The model was used to investigate global anthropogenic N and CO_2 fertilization effects on C sequestration. Modeling the historical CO_2 record, the authors found excellent agreement between GLOCO and the measured record, especially from 1850 to 1985. Their conclusion was that 40 to 70% of the total effect was attributable to N fertilization, evidence that models able to couple the C and N cycles are important for understanding the effects of climate change on terrestrial ecosystems.

Gifford et al. (1996) developed CQUEST (for "Carbon Sequestration") to investigate global C and N budgets based on atmospheric CO_2, temperature and N perturbations. They found that increasing CO_2 along with temperature and N deposition caused a net sequestration to soil and vegetation of 3 to 4 Pg-C/yr. As a globally aggregated model, CQUEST may overestimate annual values. However, it was found necessary to adjust N associated with C at below the published pool values to maintain credible C:N ratios. They concluded "that combining independent information about interlocking biogeochemical cycles can constrain the estimates of features of all of them."

Based on these models, it may be necessary to disaggregate the system and use more specific criteria to develop ecosystem C sequestration estimates. Li et al. (1994) coupled a denitrification-decomposition model and a plant growth submodel with cropping practice subroutines to investigate the effects of specific management practices on the biogeochemistry of C in arable lands. Enhanced C sequestration was found with manure or N fertilizer addition. However, the magnitude was quite sensitive to soil texture. In the long term, SOC increased with clay content, higher residue addition, and decreased temperature. Thus, these attributes are regional in scope, and specific databases for soil type, climate, soil management, and fertilizer use would be needed to construct an aggregated model for arable land. There are numerous complex interacting factors affecting SOM storage at the regional and national scales. Disaggregated models such as the Century Model (Parton et al., 1987) are adequate at predicting regional trends in SOM using inputs such as temperature, moisture, soil texture, and plant lignin content. However, at larger scales the inputs from disaggregated models become diffuse and the value for predicting SOM storage decreases.

Models attempting to account for the "missing carbon" vary depending on inputs and assumptions. However, they generally fall into a fairly narrow range from 1.0 to 4.0 Pg-C/yr. Table 1-5 lists recent estimates of annual soil and vegetation C sequestration attributable to CO_2 fertilization effects. Estimated net accumulation of terrestrial N of 50 Tg/yr produces an average C/N ratio of sequestered C of

Table 1-5. Estimates of Annual C Sequestration Resulting from Fertilization from Atmospheric CO_2

Total C Sequestration Pg/yr	Reference
1.3	Kohlmaier et al., 1989
1.4	Goudrian, 1992
2.4	Taylor and Lloyd, 1992
1.3	Polglase and Wang, 1992
2.3 (mean)	Gifford, 1994

46 (Melillo, 1996). Many models do not incorporate N cycling, and so gross C/N estimates from the models vary widely. Hudson et al. (1994) estimated that less than half of the reactive N generated per year would produce a worldwide C/N ratio of 100 for 2.3 Pg-C of sequestered C. For terrestrial biomass, these estimates may be reasonable.

Process models are useful tools for describing biogeochemical reservoirs and their interactions. However, modeling is only as reliable as the input data. For global models, it appears that gross input data inaccuracies may be smoothed by the aggregation of systems. Again, for example, Table 1-5 includes results from five studies with similar outputs for models with rather different assumptions. One consequence is that while the modeling literature is self-supportive—if a new model predicts similar values of older (or different) models, then that is validation of the model—it also contributes to the ongoing and sharp public debate. Why should we assume that a model based on C flux should return a similar value as a model that couples C and N flux or for that matter one that considers phosphorus (P)? These inconsistencies need to be addressed to help clarify the core scientific issues for policy makers and the public.

Influence of Climate Change

Increasing atmospheric CO_2 is of concern because it may drive global warming, as mentioned earlier. Both the increase in CO_2 and the resulting warming can, in turn, influence rates of C sequestration in different ways. Potentially the most significant for C storage in the terrestrial biosphere are enhanced C fixation from "CO_2 fertilization" and releases of CO_2 from northern latitude soils in response to warming.

Elevated CO_2 concentration in the ambient air probably stimulates NPP in many terrestrial ecosystems. It does this by enhancing both photosynthesis and water use efficiency. An increase in NPP predisposes an ecosystem to greater C storage through an enhanced supply of C added to standing plant biomass and litter inputs to soil C pools. The response of NPP to atmospheric CO_2 is asymptotic so the relative stimulation of NPP (and C sequestration potential) per unit increase in CO_2 declines as its concentration increases. Also, if human C sequestration efforts are successful in slowing (or eventually halting) CO_2 increase, the CO_2 fertilization response will be diminished or eliminated. Stated another way, for CO_2 fertilization to be most effective in adding C to terrestrial ecosystems, CO_2 increase will have to be rapid, which would be obvious evidence that measures enacted to stabilize its atmospheric concentration were failing (Amthor, 1998).

Warming caused by increasing CO_2 has the potential to impart a positive feedback on both CO_2 increase and warming through enhanced decomposition of soil C pools. A potential chain of events is easy to describe in general terms, but how it might play out in the real world is much less obvious. Nonetheless, the basic issue is that warming usually reduces the residence time of C in soils. This can be exacerbated if warming is accompanied by drying of wet soils. The result of the reduced residence time is an increased rate of CO_2 release from soils and other dead organic matter pools such as bogs. The increased rate of CO_2 release would increase atmospheric CO_2 concentration and reduce the effectiveness of measures enacted to enhance soil C storage. The ultimate limits to such a positive feedback are unknown, but recall that the pool of C in soils is more than twice the present atmospheric CO_2 pool. The magnitude of either (or both) the potential positive feedback on CO_2 increase through warming and enhanced decomposition or the negative feedback resulting from CO_2 fertilization will depend on the trajectories of future increases in temperature and CO_2.

NEW TECHNOLOGY FOR SOIL CARBON SEQUESTRATION

New technology with the potential to enhance soil C sequestration can be categorized as (1) technology for soil, crop and forest management; (2) exploitation of underutilized land resources and existing biodiversity; (3) plant biotechnology; (4) microbial biotechnology; and (5) chemical technology. Speculations on the potential for these new technologies to increase soil C sequestration, their likelihood of success, and the time required for their implementation have been outlined before (U.S. Department of Energy, 1997) and are further speculated upon in Table 1-6. The likelihood of success refers only to the technology being achievable; it does not consider economics or public acceptability. None of the entries in Table 1-6 consider additive potential from precision management; nor does any category include additive input from any other, although they probably overlap. All values are entirely speculative.

Precision Technology for Crop and Forest Management

As discussed above, selected management practices in agriculture and forestry have the potential to significantly enhance soil C sequestration. In addition,

Table 1-6. Proposed Global Carbon Sequestration Potential from New Technology

Technical Approach	Potential Sequestration (Pg C/yr)	Potential Success High/Med/Low (inverse of risk)	Time to Implement (yr)
Crop & Forest Management[1]			
Cropland management	0.5-2	H	Now
Forest management	1-3	M/H	Now
Underutilized Resources			
Deserts & saline/alkaline aquifers[2]	0-1	L	25
Native plant biodiversity[3]	N/A	H	Now
Plant Biotechnology[4]			
Innate productivity	1-6	L/M	25
Photoassimilate partitioning	0-2	L/M	25
Stress tolerance (salt, drought)	1-3	M	10
Nitrogen fixation	1-2	L/M	25
Lignin (fiber) biosynthesis	1-2	L/M	5
Bioproducts[5]	N/A	H	5
Microbial Biotechnology			
Soil conditioners	0-0.5	M	Now
Mycorrhizal fungi	1-2	H	10
Engineering rhizobial communities	1-2	H	25
Chemical Technology			
Smart fertilizers	0-1	M	10
Soil additives/conditioners	0-0.5	H	Now
Calcium for arid regions	0-2	L	Now
Plant growth regulators	0-1	H	Now

[1] Does not include impacts of biotechnology or chemical technology, which are probably additive.
[2] Baseline is unmanaged arid/hyper-arid desert without plant cover. Lower value is periodic desert flooding. Larger value is for highly engineered microalgal mass culture.
[3] Because many scenarios are possible, there is no single comparison base.
[4] All of the plant biotechnology options are intimately linked, so values are probably not additive.
[5] Fossil fuel offsets.

sequestration potential could be further enhanced with new technology. Ground- and plant-based sensing mechanisms linked to proximal or remote imaging with computer control comprise the emerging set of technologies known as precision agriculture (Pierce and Sadler, 1997). In time, the widespread use of precision methods over broad areas of agriculture and silviculture holds great potential for use in efforts to enhance soil C sequestration. The contribution of this "high" technology to precision farming and forestry will include the following:

- Sensors to detect the appearance of pathogens for early, precisely targeted control
- Use of aerial and global positioning technology for precise application of fertilizer—how much, when and where it is needed in a field or forest
- Efficient "just in time" irrigation systems that maximize water use efficiency.

Each of these approaches will apply rapidly evolving computer and information technology. In the near term, sensors will be electronic devices. In the longer term, plants may be genetically engineered to produce unique spectral signals in response to stress such as pathogen attack or water deficit, for detection by novel and cost-effective remote sensing systems (e.g., hyperspectral imaging) now in the early stages of development.

Underutilized Resources

Underutilized resources include land, groundwater, and native biodiversity. For land and groundwater, the related cases considered here are (1) to irrigate deserts with water from saline and alkaline aquifers, or (2) to use the saline/alkaline groundwater to mass culture microalgae. Microalgae, including cyanobacteria (blue-green algae), include species that are natural components of biological crusts in semiarid lands and deserts and capable of fixing C and N at high temperature and irradiance when provided with sufficient moisture (Metting, 1991). Although never attempted, it is theoretically possible to use periodic flooding with saline/alkaline groundwater to encourage the growth of salt and alkaline-tolerant cyanobacterial inoculants over large areas of desert (Knutsen and Metting, 1991). In addition to photosynthesis, enhanced C sequestration through carbonate formation might also take place if the groundwater were rich in Ca.[9] Use of the same groundwater resources to mass culture microalgae in highly engineered systems in deserts has been the subject of ongoing research for over 25 years. Although microalgae are cultivated commercially for high value products (Metting, 1996), the large-scale production of microalgae strictly for biomass or C capture has yet to be attempted because of unfavorable economics. Because the single largest projected cost is the acquisition and provision of CO_2, it may be feasible to transport sequestered C from power plants to deserts for this purpose (Benemann et al., 1984).

---

[9]Any scheme of this nature, however, would have to carefully consider the potential loss of CO_2 from carbonate dissolution depending on site-specific parent materials and geochemical conditions (Schlesinger, personal communication).

The more practical near-term opportunity is to exploit underutilized native plant resources with, for example, inherent tolerance to acid soil, salinity, drought or other stresses for purposes such as reclamation of degraded land and erosion control to stem desertification. This might be accomplished with a focus on selected C-3, C-4, and CAM[10] plants for water use efficiency and by expanding the area planted to shelter belts and to food and biomass crops shown to be effective under various ecological situations. In this volume Lal et al. consider in detail the use of native plant biodiversity for restoring degraded lands.

Plant Biotechnology

Plant genetic engineering for enhanced NPP could lead directly to greater C sequestration. Changes in traits other than NPP, such as chemical composition or partitioning between shoot and root, could also influence C sequestration. Some targets endpoints for enhanced C sequestration via plant biotechnology are to:

- Improve innate photosynthetic efficiency (NPP)
- Manipulate photoassimilate partitioning
- Manipulate content of lignin and other polymers
- Develop various bio-products to displace fossil fuel based products
- Improve stress tolerance, including salt and drought tolerance, and aluminum (Al) tolerance for acidic tropical soils
- Engineer C-4 photosynthesis into C-3 plants
- Engineer N_2-fixation into non-leguminous plants.

Two general points emerge from such a list. First, the application of most or all of the technologies would likely be limited to highly managed ecosystems, such as croplands, managed forests and biomass plantations. In addition to significant logistical challenges, the introduction of genetically altered characteristics into native plant communities and ecosystems could engender societal resistance based on environmental and ethical concerns. Second, many of the proposed modifications have been the subject of research for decades, with limited success. Notable exceptions include altered photoassimilate partitioning in favor of seeds in grain crops and directed reductions (rather than increases) in lignin content of silage. Together, these experiences imply that large-scale changes beyond agriculture and silviculture and the chances for large changes in key plant characteristics may be limited. This may largely be because many of the proposed changes to plant processes involve multiple genes and the coordination of biochemistry with anatomy, as well as complex and interacting metabolic pathways (ap Rees, 1995).

---

[10]CAM = crassulacean acid metabolism. Like C-4 species, CAM plants combine a specialized leaf anatomy and biochemistry for CO_2 fixation.

This suggests that directed genetic solutions to complex traits will be difficult, much more so than changes involving only one (or a few) genes. Against that backdrop, we appraise each target briefly.

Innate photosynthesis and NPP might be improved in C-3 species if the characteristics of ribulose 1,5-bisphosphate carboxylase/oxygenase (rubisco) could be modified. An obvious target is the CO_2/O_2 specificity of this bifunctional enzyme. If the specificity of rubisco for CO_2 could be improved, net photosynthesis would increase by reducing or eliminating photorespiration. A simple measure of the potential for increased photosynthesis is the known ratio of photorespiration to photosynthesis. With present day atmospheric CO_2 and O_2 concentrations, the photorespiration/photosynthesis ratio varies from 0.1 to 0.3 for most C-3 plants (Amthor, 1995). Theoretically, photosynthesis could thus be increased 10 to 30% if photorespiration were to be eliminated by engineering rubisco. Although molecular approaches to modifying plant physiology are advancing, decades of previous research have contributed little improvement to photosynthesis *per se*, except as associated with canopy architecture and plant nutrition. Of course, increased photosynthetic capacity would have to be coupled with increased "sink" capacity to obtain greater NPP.

For purposes of soil C sequestration, a key aspect of partitioning is the relative allocation to roots. Thus, modification and control of root architecture and rooting depth is an important target whose realization depends on improved understanding of underlying genetic and metabolic processes (Schiefelbein et al., 1997). Crop breeders have increased harvest index (ratio of grain to total biomass) in cereals, but the accompanying changes in the ratio of above-ground to below-ground growth have varied (Evans, 1993). If increased root growth comes at the expense of grain or other harvest targets, farmers would likely be hesitant to adopt cultivars with these characteristics.

Modification of the lignin content and structure of plants is another potential biotechnology for enhancing C sequestration (Lewis and Yamamoto, 1990). Vascular plants may contain up to 20% of their dry weight in lignified C compounds (Whetten et al., 1998). These highly aromatic materials provide mechanical support, transport water, and defend against pathogen attack. The insolubility and complexity of lignin polymers render them resistant to degradation by most microorganisms, making them more persistent in soils than is true for cellulose and other non-aromatic compounds (Campbell and Sederhoff, 1996). There are several ongoing efforts to modify lignin content which primarily seek to reduce content for increased ruminant digestibility and higher quality paper pulp (Sewalt et al., 1997). The basic knowledge gained from these efforts to control the synthetic pathways could perhaps be directed to the opposite purpose of increasing lignin production, for example, in the roots of select grasses. It is important to realize that the quality of lignin varies among plant species. Also, fungi and other microorganisms have evolved to use lignin as a C source so that enhanced microbial degradation may offset increased lignin input to soil.

Any genetically based improvements in productivity of biofuel crops have the potential to reduce fossil fuel use. While this is not a direct sequestration strategy, any reduction in fossil fuel use reduces the total requirement for C sequestration. That is, the biofuel production is sustainable and does not result in ecosystem C losses greater than the fossil fuel savings, such as might occur when an existing ecosystem is modified for a biofuel plantation.

Improved stress tolerance would result in greater NPP and could also increase the distribution of plant species. Key stresses relevant to enhancing C sequestration include drought and salinity, which affect vast areas of arid and semi-arid ecosystems; Al stress in tropical acid soils is also of considerable importance and areal extent. As with other complex ecological interactions, stress tolerance is generally determined by multiple biochemical pathways and signaling networks that control acquisition of water and nutrients, chloroplast function, and synthesis of stress proteins and osmotically-active metabolites (Koch, 1996). To date, the transfer of individual genes has resulted in only marginal enhancement of stress tolerance, demonstrating again the necessity for understanding multiple gene interaction (Bohnert and Jensen, 1996). The benefits in terms of enhanced NPP and potential C sequestration that could be associated with improved stress tolerance justify a concerted research effort in this area.

Net primary production by plants with C-4 photosynthetic metabolism, most notably tropical grasses, can exceed that of C-3 plants, at least in environments to which they are adapted. Therefore, replacement of C-3 with C-4 physiology in select species might be an avenue toward enhanced C sequestration in terrestrial ecosystems. Importantly, even though C-4 plants account for no more than 1% of higher-plant species, they may already account for 20-25% of global photosynthesis in non-crop ecosystems (Lloyd and Farquhar, 1994), so C-4 photosynthesis is already "active" in the terrestrial C cycle.

The notion of C-4 wheat (by nature a C-3 species) has been of interest to crop scientists for some time (Dai et al., 1993). Unfortunately, C-4 photosynthesis is associated with an integrated complex of biochemical and anatomical features that are not easily introduced into C-3 plants, even though C-4 photosynthesis/anatomy has apparently evolved on several separate occasions (Ehleringer and Monson, 1993; Furbank and Taylor, 1995). Photosynthesis in C-4 plants depends on unique leaf anatomy, so that "inserting" C-4 photosynthesis into C-3 plants using biotechnology has not yet been achieved, although a recent study reported successful expression of a C-4 enzyme in rice, a C-3 plant (Ku et al., 1998). Also, C-4 photosynthesis can be inferior to C-3 photosynthesis in cold environments, so the utility of C-4 photosynthesis can be limited geographically. Moreover, woody plants, which almost universally lack C-4 metabolism, may be more useful for storing C over time periods of a few decades. This too indicates a limited prospect for sequestering large amounts of C globally by engineering a shift from C-3 to C-4 photosynthesis—unless it could be inserted into forests, which seems unlikely outside the scope of tree plantations.

Another long-sought approach to greatly improving the efficiency and reducing the cost of crop production is by engineering N_2-fixation into plant species (particularly grasses) other than legumes and the few other plant families that harbor bacterial N_2-fixing symbionts. The minimal progress toward introducing symbiotic

N_2-fixation into non-legumes again probably reflects an inherent limitation for directed modification of non-legumes due to the complex, multigenic nature of the symbiosis.

Microbial Biotechnology

Microbiological approaches to enhanced C sequestration are also possible. As with plants, near-term opportunities include screening and selection of existing biodiversity to improve microbial-plant symbioses, such as mycorrhizal fungi, bacterial N_2-fixation, biological control, phosphate solubilization, and soil conditioning (Metting, 1992). Developing the capability to grow mycorrhizal fungi (and other unculturable microorganisms) in pure culture would be tremendously valuable. These fungi are symbiotic with all important forest tree species and most crops, for which they enhance water gathering efficiency and nutrient (P and microelements) uptake (Gianinazzi-Pearson, 1996). Genetic engineering to improve these traits, as well as enhancing the energy efficiency of N_2-fixation in free-living and symbiotic bacteria, is another mid- to long-term (10-25 yr) research target. Research has begun with coupled plant-microbial systems with the aim of manipulating entire microbial rhizosphere communities by modulating root exudate patterns (Savka and Farrand, 1997). In order to achieve these objectives, it will be necessary to first understand the intricacies of symbiosis and the very complex ecology of the rhizosphere (Long, 1996; Weller and Thomashow, 1994).

Finally, microbial production of polysaccharides and humic materials facilitates formation and stabilization of aggregates (Lynch and Bragg, 1985). Microbial inoculants for conditioning soil have been developed and used on a limited scale (Knutsen and Metting, 1991; Metting, 1996).

Chemical Technology

Other potential new technology for enhanced soil C sequestration includes novel fertilizers and soil amendments based on advances in materials research, and the use of plant growth regulators (PGRs). "Smart" fertilizers might be developed to release nutrients upon demand by the plant in response to specific molecular signals. Chemical products (e.g., polyvinyl alcohol, lignites) to promote and stabilize soil aggregation have been introduced for erosion control in the past, but have not been economical for large-scale use in agriculture (Armbrust and Dickerson, 1971).

For carbonates to act as a sink for atmospheric CO_2, a source of Ca is required. A net transfer of C to the soil does not take place if the Ca is derived to begin with from dissolved carbonate. This is particularly important for alkaline soils in arid and semi-arid regions where increased irrigation or precipitation without Ca input could result in net emissions of CO_2 from carbonate dissolution (Grossman et al., 1995). If economical sources of Ca are identified, the potential exists to enhance carbonate formation in alkaline soils by means of Ca addition. There may also exist microbial processes in soil similar to those in aquatic and marine systems for carbonate formation, but these have not been extensively studied.

Plant growth regulators (PGRs) are synthetic and natural compounds (mostly organic), including hormones, that affect plant metabolism and development when present in very low concentrations. Many PGRs have been used commercially for many years to hasten or delay the ripening of fruit, to induce senescence, increase production of a desired metabolite (e.g., latex, oleoresins), or control plant morphology, including tillering in grasses (Johnson, 1982). Molecular signals from symbiotic and pathogenic soil microorganisms are also known to influence root architecture when present in low concentrations in the rhizosphere. Thus, the discovery, selection, and use of select PGRs to influence below-ground growth and metabolism may also be a feasible approach to enhance soil C sequestration.

KNOWLEDGE GAPS AND SCIENTIFIC RESEARCH NEEDS

The Overarching Questions

The existence of many fundamental knowledge gaps highlights the need for scientific research to improve our understanding of soil C sequestration and establish an information base for the development and implementation of new sequestration technology. Ongoing and future research efforts and the identification of specific gaps in basic understanding with which they are associated can be addressed in the context of a set of overarching questions in need of resolution:

- How can we best reduce the large uncertainties in global terrestrial C inventories?
- Are estimated native ecosystem C sequestration capacities equivalent to their maximum carrying capacities? Is the historic (pre-agriculture) storage capacity of agricultural soils equivalent to their maximum inherent capacity? How accurate an estimate of the long-term potential for soil C sequestration can be made?
- Will tropical and northern temperate forests, and high latitude taiga and tundra ecosystems become net sinks or sources of C with global warming? Will warming favor enhanced production or SOM oxidation? How will this vary among regions?
- How can other potential benefits and risks of enhanced soil C sequestration, such as soil quality (e.g., Doran et al., 1994), best be identified and quantified?
- What is the potential for plant and microbial genetic engineering in the post-genome era to make significant impacts on C sequestration? What are the inherent biological and ecological limits to NPP and to plant-microbe symbiosis?
- What are the opportunities for developing new biological, chemical, and C sequestration management technologies? Which of these can realistically be applied over large areas of the Earth?

Soil Carbon Inventories: Database Consistency and Verification

More accurate baseline inventories of global land use, extent and sequestration capacity of native and disturbed ecosystems, and stocks of organic and inorganic soil C are needed. To be useful for long-term forecasting and planning, it is important to have not only more reliable present day inventories, but also better inventories of C stocks prior to and during expansion of agriculture in the nineteenth century. Current inventories for some regions, estimates of historic CO_2 release to the atmosphere, and projected sizes of future terrestrial C sinks often carry uncertainties of 25-50% or more. This results from imperfect assumptions and the use of data collected mostly for other purposes. These data are statistically inadequate for representing large areas of the Earth's land surface. For example, it has been speculated that the "missing" C sink may just be a result of misinterpretation of inadequate databases (Lal et al., 1998c).

Baseline data and improved inventories that are statistically significant and consistent over large areas of the Earth are needed. Specifically, studies designed to collect information on the scale of kilometers or smaller are needed for large areas of Africa, South America, and Asia because existing data are not accurate enough for predictive and interpretive purposes. Also, special attention should be paid to categorization and prioritization of degraded lands for restoration and of saline, alkaline, and acid sulfate soils to enact programs to head off future degradation (Scharpenseel et al., 1990).

Data collection should also be designed in such a way as to improve our ability to project influences on soil C of land use changes such as, for example, urbanization, deforestation, afforestation and reforestation, or the creation of biofuel plantations. Information is also needed on a number of selected ecosystem parameters that are not now commonly collected in order to better understand soil C sequestration. In addition to above-ground biomass and SOC data, efforts should include data on litter and coarse woody debris, neither of which is routinely included in commercial forest inventories. Also, it is very important to quantify SIC, for which native stocks and dynamics and relationships to land use are virtually unknown. An example of collaborative efforts to improve databases is SOMNET, the International Soil Organic Matter Network.

Soil Carbon Dynamics and Pedospheric Processes

Despite recent progress toward improved national and regional soil C budgets, research and modeling of soil C dynamics, and research in land use and soil management, many knowledge gaps still remain in our understanding of the fundamental mechanisms responsible for soil C sequestration and interrelated pedospheric processes (Lal et al.,1997). Quantitative data on relative C turnover rates in macro- versus microaggregates are lacking (Jastrow et al., 1996). Also, very few data are available on below-ground vegetation C stocks, which are essential to understanding soil C allocation and flux.

The proportion of C stocks below the active decomposition zone is important to long-term storage. What processes control the downward movement of C? The production and fate of dissolved organic C (DOC) is poorly understood, as is its potential for leaching and deep storage. Recent discoveries of large and pervasive subsurface microbial ecosystems (Fredrickson and Onstott, 1995) suggest that biogeochemical mechanisms may exist at the deep root/subsoil interface to control the fate of C at depth. Research in this area is warranted.

Also in need of better understanding are the mechanisms responsible for cycling and allocation of soil C, including detrital chemistry and SOM formation, and formation and stabilization of soil micro- and macroaggregates and their role in sequestration. Soils in which it is particularly important to focus new efforts are those from tropical ecosystems, in which both soil C losses and gains are rapid at high temperature. Equally in need of additional research are (1) C pools in frozen soils and the potential for changes in C dynamics, (2) the magnitude and dynamics of SIC, in both arid and non-arid regions, and (3) C sequestration in subsoils. Of particular importance is the need for better understanding on regional and global scales of interactions among C, water, and other major biogeochemical cycles [N, sulfur (S), P, iron (Fe)] and how human activities impact these cycles.

Pedospheric processes important to regional and global C cycling include leaching, erosion, and gaseous flux (Lal et al., 1997). The fate of C redistributed over the landscape by erosion and through the profile by leaching is poorly understood. One example of the possible consequences of this lack of understanding is the inability to predict the scale of potential impacts on pedospheric processes of fertilizer use and intensive agriculture in ecoregions not previously subjected to large-scale farming, such as tropical savannahs.

Potential directions and rates of change of important edaphic properties and processes with changing land use and climate will also be easier to predict with improved fundamental understanding (Scharpenseel et al., 1990). Research is needed to bound the probable time scales for pedospheric changes and their probable effects on C sequestration. The influences of ongoing and expected global climate change on soil processes include increased temperature, changing moisture patterns, and increased atmospheric CO_2. Important questions include:

- What processes govern rates of change in soil salinity and alkalinity on time scales from months to decades?
- What key activities of soil microbiota, microfauna, and mesofauna are central to C sequestration in months, years, and decades? How are these reflected in soil productivity and quality?
- How do stocks and the quality of C evolve in different soils over years and decades, and how are they influenced by initial structural, textural, and mineralogical properties?
- How will the susceptibility of soil to erosive forces change with addition or loss of C? What will be the time scale of these changes—years or decades?

Biotechnology and Biodiversity

Research needs in plant and microbial biodiversity center on discovery and field testing. Existing programs and technical approaches to identify novel biodiversity can readily be redirected for purposes of soil C sequestration. The principal limitation is funding for these efforts. Lal et al. (in this volume) examine in detail the research needs for discovery and application of native plant biodiversity for C sequestration and reclamation of degraded lands.

Research in microbial diversity and application is currently limited because most microorganisms are unculturable. Direct extraction and study of DNA and RNA from the environment (Chandler, 1998) and the rapid growth in available whole microbial genome sequences (Kyrpides)[11] are together opening whole new avenues to understanding and applying microbial biology. Nonetheless, the improved ability to culture important microorganisms, such as mycorrhizal fungi, remains key to advancing microbial applications to enhanced soil C sequestration.

As mentioned, many or most of the properties of plants that are targets for enhanced sequestration are multigenic in nature, involving numerous metabolic and molecular signaling events and pathways. Building on genome sequencing, research in functional genomics and proteomics (protein diversity and function) is needed to unravel the complexities of multigenic traits and their interaction with the environment before directed modification can proceed effectively. Thus, the use of ever more sophisticated molecular and instrumental approaches is important to direct basic research at key biological and ecological processes, such as plant-microbe molecular signaling, relatedness of environmental stress to plant NPP, and biological N_2-fixation, to name but three examples.

Methods and Instrumentation

Much of the needed research, such as improving soil C inventories, can be addressed by broader use of currently available methods. In other cases, new approaches to integrated field and laboratory research and better instrumentation are required. The parallel improvement of process-oriented computational models for simulating key mechanisms at the appropriate molecular, pore, and meter scales, and relating them to field and ecosystem scales is also important for understanding and predicting the potential consequences to soil C sequestration of shifting land use and global climate change (Paustian et al., 1995). Experimental approaches for addressing field research needs include:

- Analog studies with soil chronosequences
- Improvement of process models for describing changes in soil properties as a function of temperature, precipitation, and land use
- Field manipulation studies
- Long-term field studies.

[11]N. Kyripides. "Complete and ongoing genome projects—microbes." http://geta.life.uiuc.edu/~nikos/genomes.html

Priority ecosystems for field research to understand potential impacts of global climate change include transitional regions, coastal areas, irrigated landscapes in semiarid regions, tundra, deforested areas, and natural wetlands (Scharpenseel et al., 1990). Studies to understand the possible impacts of long-term tradeoffs between enhanced C sequestration and reduction in gaseous C emissions from soil are needed to address the question of how different ecosystems will shift or change (Houghton et al., 1998). For example:

- Minimum tillage slows the rate of decomposition of SOC yet adds organic materials to the surface. Is occasional cultivation to bury the additional organic matter required to maximize long-term C sequestration?
- What will be the effects of expanded use of N and other nutrient fertilization, such as in forests? Productivity will increase, but will SOM increase or will enhanced microbial activity lead to release of C to the atmosphere?

Finally, new analytical methods and instruments (e.g., NMR and mass spectrometers) currently under development should be employed to improve and standardize data on soil C. Analytical methods must be standardized and widely adopted for accurate and precise quantification of different C pools while reproducibly and significantly detecting the *very* small concentrations and changes in soil C that are important on regional and global spatial scales over years, decades, and longer. Advanced isotopic and spectroscopic methods should be improved upon and used on a larger scale. Development of reliable instruments for routine measurement of SOC and SIC in the field is also an important goal.

Socioeconomic Considerations

Research is needed to understand the overall social and economic advantages and challenges of applying new technology to enhance soil C sequestration. Central to this is the requirement for complete life cycle inventory and environmental analysis for competing land use options and for the introduction of new technologies for soil C sequestration.

SUMMARY

Storage of C in soils has the potential to offset CO_2 emissions to the atmosphere in the coming decades while new "clean" energy production and CO_2 sequestration technologies are developed and deployed. Two-thirds of the C in the terrestrial biosphere is stored below-ground. Its worldwide distribution as SOM and SIC among ecosystems is not accurately known. Also poorly understood are the biogeochemical and pedogenic mechanisms responsible for C allocation into pools of varying longevity and how the key processes are manifested at the molecular, microscopic, soil aggregate, field, landscape, regional and global scales.

Because they are economically important and have a rich history of directed research, forests and croplands are best suited for application of existing and new technology to enhance soil C sequestration in the near term. Nonetheless, the estimated potential for enhanced C storage, even in the United States, varies more than two-fold. And beyond current management approaches, new management, chemical, and biological technology has the potential to impact soil C storage. But we can only speculate about the extent of the impact. What is needed is, first, basic research to improve our fundamental understanding of natural phenomena controlling soil C sequestration and, second, basic and applied research and development to bring new management and technology to the challenge.

REFERENCES

Amthor, J.S. 1995. Terrestrial higher plant response to increasing atmospheric (CO_2) in relation to the global carbon cycle. *Global Change Biology* **1**:243-274.

Amthor, J.S. 1998. Perspective on the relative insignificance of increasing atmospheric CO_2 concentration to crop yield. *Field Crop Research* **58**:109-127.

Amthor, J.S., and M.A. Huston (eds). 1998. *Terrestrial Ecosystem Responses to Global Change. A Research Strategy*. Oak Ridge National Laboratory, ORNL/TM-1998/27, Oak Ridge, TN.

ap Rees, T. 1995. Prospects of manipulating plant metabolism. *Trends in Biotechnol.* **13**:375-387.

Armbrust, D.V., and J.D. Dickerson. 1971. Temporary wind erosion control: Cost and effectiveness of 34 commercial materials. *J. Soil Water Conservation* **26**:154-157.

Armentano, T.V., and E.S. Menges. 1986. Patterns of change in the carbon balance of organic soil-wetlands of the temperate zone. *Journal of Ecology* **74**:755-774.

Benemann, J.R., J.C. Weissman, R.P. Goebel, and D.C. Augenstein. 1984. Microalgae fuel economics. *Nova Hedwigia* **83**:176-191.

Benson, S., W. Chandler, J. Edmonds, M. Levine, L. Bates, H. Chum, J. Dooley, D. Grether, J. Houghton, J. Logan, G. Wiltsee, and L. Wright. 1997. *Carbon Management: Assessment of Fundamental Research Needs*. U.S. Department of Energy, DOE/ER-0724.

Birdsey, R.A. 1992. Carbon storage in trees and forests. In R.N. Sampson and D. Hair (eds), *Forests and Global Change. Vol. 2.* American Forests. Washington, DC: 23-33.

Bohnert, H.J., and R.G. Jensen. 1996. Strategies for engineering water-stress tolerance in plants. *Trends in Biotechnol.* **14**:89-91.

Bouwman, A.F. (ed). 1990. *Soils and the Greenhouse Effect*. John Wiley & Sons, New York.

Burke, I.C., W.K. Lauenroth, and D. G. Milchunas. 1997. Biogeochemistry of managed grasslands in central North America. In E.A. Paul et al. (eds), *Soil Organic Matter in Temperate Agroecosystems*. CRC Press, Boca Raton: 85-102.

Campbell, M.M., and R.R. Sederhoff. 1996. Variation in lignin content and composition. *Plant Physiol.* **110**:3-13.

Chandler, D.P. 1998. Redefining relativity: quantitative PCR at low template concentrations for industrial and environmental microbiology. *J. Industrial Microbiology* **21**:128-140.

Cole, V., C. Cerri, K. Minami, A. Mosier, N. Rosenberg, and D. Sauerbeck. 1996. Agricultural options for mitigation of greenhouse gas emissions. In R.T. Watson, M.C. Zinyowera, and R.H. Moss (eds), *Climate Change 1995: Impacts, Adaptations, and Mitigation of Climate Change: Scientific-Technical Analyses.* Contribution of Working Group II to the Second Assessment Report of the Intergovernmental Panel on Climate Change. Cambridge University Press, Cambridge:744-771.

Coupland, R.T. 1992. Mixed prairie. In R.T. Coupland (ed), *Natural grasslands: introduction and western hemisphere.* Elsevier, New York: 151-182.

Dai, A., and I.Y. Fung. 1993. Can Climate Variability Contribute to the "Missing" CO_2 Sink? *Global Biogeochemical Cycles* **7**:599-609.

Dai Z., M.S.B. Ku, and G. Edwards. 1993. C-4 photosynthesis. The CO_2 –concentration mechanism and photorespiration. *Plant Physiology* **103**:83-90.

Dixon, R.K., S. Brown, R.A. Houghton, A.M. Soloman, M.C. Trexler, and J. Wisniewski. 1994. Carbon pools and flux of global forest ecosystems. *Science* **263**:185-190.

Doran, J.W., D.C. Coleman, D.F. Bezdicek, and B.A. Stewart (eds). 1994. *Defining Soil Quality for a Sustainable Environment.* Soil Science Society of America Special Publication Number 35, Madison, WI.

Eddy, J.A., and H. Oeschger (eds). 1995. *Global Changes in the Perspective of the Past.* John Wiley, Chichester, UK.

Ehleringer, J. R., and R. K. Monson. 1993. Evolutionary and ecological aspects of photosynthetic pathway variation. *Annual Review of Ecology and Systematics* 24:411-439.

Eswaran, H., E. Van den Berg, P. Reich, and J. Kimble. 1995. Global soil carbon resources. In Lal, R., J. Kimble, E. Levine, and B.A. Stewart (eds), *Soils and Global Change.* CRC Lewis Publishers, Boca Raton: 27-43.

Evans, L. T. 1993. *Crop Evolution, Adaptation and Yield.* Cambridge University Press, Cambridge.

Fan, S., M. Gloor, J. Mahlman, S. Pacala, J. Sarmiento, T. Takahashi, and P. Tans. 1998. A large terrestrial carbon sink in North America implied by atmospheric and oceanic carbon dioxide data and models. *Science* **282**:442-446.

Fredrickson, J.K., and T.C. Onstott. 1995. Microbes deep inside the Earth. *Scientific American* **275**:68-73.

Furbank, R.T., and W.C. Taylor. 1995. Regulation of photosynthesis in C-3 and C-4 plants: a molecular approach. *The Plant Cell* **7**:797-807.

Gianinazzi-Pearson, V. 1996. Plant cell responses to arbuscular mycorrhizal fungi: Getting to the roots of the symbiosis. *The Plant Cell* **8**:1871-1883.

Gifford, R.M. 1994. The global carbon cycle: A viewpoint on the missing sink. *Australian Journal of Plant Physiology* 21:1-15.

Gifford, R.M., J.L. Lutze, and D. Barrett. 1996. Global atmospheric change effects on terrestrial carbon sequestration: Exploration with a global C- and N-cycle model (CQUESTN). *Plant and Soil* **187**:369-387.

Goudriaan, J. 1992. Biosphere structure, carbon sequestering potential and the atmospheric ^{14}C carbon record. *Journal of Experimental Botany* **43**:1111-1119.

Groot, J. J. R., P. de Willigen, and E. L. J. Verberne (eds). 1991. *Nitrogen Turnover in the Soil-Crop System*. Kluwer Academic Publishers, Dordrecht, The Netherlands.

Grossman, R.B., R.J. Ahrens, L.H. Gile, C.E. Montoya, and O.A. Charwick. 1995. Areal evaluation of organic and carbonate carbon in a desert area of southern New Mexico. In Lal, R., J. Kimble, E. Levine, and B.A. Stewart (eds), *Soils and Global Change*. CRC Lewis Publishers, Boca Raton: 81-91.

Hair, D., R.N. Sampson, and T.E. Hamilton. 1996. Summary: forest management opportunities for increasing carbon storage. In R.N. Sampson and D. Hair (eds), *Forests and Global Change: Vol 2, Forest management opportunities for mitigating carbon emissions*. American Forests, Washington DC: 237-254.

Harrison, K., and W. Broecker. 1993. A strategy for estimating the impact of CO_2 fertilization on soil carbon storage. *Global Biogeochemical Cycles* 7:69-80.

Houghton, R.A. 1995. Changes in the storage of terrestrial carbon since 1850. In Lal, R., J. Kimble, E. Levine, and B.A. Stewart (eds), *Soils and Global Change*. CRC Lewis Publishers, Boca Raton: 45-65.

Houghton, R.A., E.A. Davidson, and G.M. Woodwell. 1998. Missing sinks, feedbacks, and understanding the role of terrestrial ecosystems in the global carbon balance. *Global Biogeochemical Cycles* **12**:25-34.

Houghton, J.T., L.G. Meira Filho, B.A. Callendar, N. Harris, A. Kattenberg, and K. Maskell (eds). 1996. *Climate Change 1995: The Science of Climate Change*. Report of the Intergovernmental Panel on Climate Change. Cambridge University Press, Cambridge, England.

Hudson, R.J.M., S.A. Gherini, and R.A. Goldstein. 1994. Modeling the global carbon cycle: Nitrogen fertilization of the terrestrial biosphere and the "missing" CO_2 sink. *Global Biogeochemical Cycles* **8**:307-33.

Hyvönen, R., G.I. Ågren, and E. Bosatta. 1998. Predicting long-term soil carbon storage from short-term information. *Soil Science Society of America Journal* 62:1000-1005.

Jastrow, J.D., T.W. Boutton, and R.M. Miller. 1996. Carbon dynamics of aggregates-associated organic matter estimated by carbon-13 natural abundance. *Soil Sci. Soc. Amer J.* **60**:801-807.

Johnson, J.C. 1982. *Plant Growth Regulators and Herbicide Antagonists*. Chemical Technology Review No. 212. Noyes Data Corporation, Park Ridge, NJ.

Kay, B.D. 1998. Soil structure and organic carbon: A review. In Lal, R., J. Kimble, R.F. Follett, and B.A. Stewart (eds), *Soil Processes and the Carbon Cycle.* CRC Lewis Publishers, Boca Raton: 169-197.

Kimble, J.M., H. Eswaran, and T. Cook. 1990. Organic carbon on a volume basis in tropical and temperate soils. *14th International Soil Science Congress* **5**:248-258. Kyoto, Japan.

Knutsen, G., and F.B. Metting. 1991. Microalgal mass culture and forced development of biological crusts in arid lands. In J. Skujins (ed), *Semiarid Lands and Deserts. Soil Resource and Reclamation*. Marcel Dekker, New York: 487-506.

Koch, K.E. 1996. Carbohydrate-modulated gene expression in plants. *Ann. Rev. Plant Physiology Plant Molecular Biol.* **47**:509-540.

Kohlmaier, G.H., E. O. Sire, A. Janecek, C. D. Keeling, S. C. Piper, and R. Revelle. 1989. Modeling the seasonal contribution of a carbon dioxide fertilization effect of the terrestrial vegetation to the amplitude increase in atmospheric carbon dioxide at Mauna Loa Observatory, Hawaii, USA. *Tellus Series B Chemical And Physical Meteorology* **41**:487-510.

Ku, M.S.B., A. Sakae, M. Nomura, S. Hirose, S. Toki, and M. Miyao-Tokutomi. 1998. High level expression of maize phosphoenolpyruvate carboxylase in transgenic rice plants. *Nature Biotechnology* **17**:76-80.

Kucera, C.L. 1992. Tall-grass prairie. In R.T. Coupland (ed), *Natural grasslands: introduction and western hemisphere.* Elsevier, New York: 227-268.

Lal, R. 1990. Soil erosion and land degradation: The global risks. *Advances in Soil Science* **11**:129-172.

Lal, R., H.M. Hassan, and J. Dumanski. 1999. Desertification control to sequester C and mitigate the greenhouse effect. In *Carbon Sequestration in Soils: Science, Monitoring and Beyond*, N.J. Rosenberg, R.C. Izaurralde, and E.L. Malone (eds). Battelle Press, Columbus, OH.

Lal, R., J. Kimble, E. Levine, and B.A. Stewart (eds). 1995a. *Soils and Global Change.* CRC/Lewis Publ., Boca Raton.

Lal, R., J. Kimble, E. Levine, and B.A. Stewart (eds). 1995b. *Soil Management and Greenhouse Effect.* CRC/Lewis Publishers, Boca Raton.

Lal, R., J. Kimble, and R. Follett. 1997. Knowledge gaps and researchable priorities. In Lal, R., J. Kimble, R.F. Follett, and B.A. Stewart (eds*), Soil Processes and the Carbon Cycle.* CRC Lewis Publishers, Boca Raton: 595-604.

Lal, R., J. Kimble, and R. Follett. 1998a. Land use and soil C pools in terrestrial ecosystems. In Lal, R., J. Kimble, R.F. Follett, and B.A. Stewart (eds), *Management of Carbon Sequestration in Soil*. CRC Lewis Publishers, Boca Raton: 1-10.

Lal, R., J. Kimble, V.C. Cole, and R. Follett. 1998b. *The Potential of U.S. Cropland to Sequester Carbon and Mitigate the Greenhouse Effect.* Ann Arbor Press, Ann Arbor.

Lal, R., J. Kimble, and R. Follett. 1998c. Need for research and need for action. In Lal, R., J. Kimble, R.F. Follett, and B.A. Stewart (eds*), Management of Carbon Sequestration in Soil*. CRC Lewis Publishers, Boca Raton: 447-454.

Lauenroth, W.K., and D.G. Milchunas. 1992. Short-grass steppe. In R.T. Coupland (ed), *Natural grasslands: introduction and western hemisphere.* Elsevier, New York: 183-226.

Lewis, N.G., and E. Yamamoto. 1990. Lignin: occurrence biogenesis and biodegradation. *Ann. Rev. Plant Physiol.* **41**:455-496.

Li, C., S. Frolking, and R.C. Harriss. 1994. Modeling carbon biogeochemistry in agricultural soils. *Global Biogeochemical Cycles* **8**:237-54.

Lloyd, J., and G.D. Farquhar. 1994. 13C discrimination during CO_2 assimilation by the terrestrial biosphere. *Oecologia* **99**:201-215.

Long, S.R. 1996. *Rhizobium* symbiosis: Nod factors in perspective. *The Plant Cell* **8**:1885-1889.

Lynch, J.M., and E. Bragg. 1985. Microorganisms and soil aggregate stability. *Advances in Soil Science* **2**:133-171.

McGuire, A.D., J.M. Melillo, L.A. Joyce, D.W. Kicklighter, A.L. Grace, B. Moore III, and C.J. Vorosmarty. 1992. Interactions between carbon and nitrogen dynamics in estimating net primary productivity for potential vegetation in North America. *Global Biogeochemical Cycles* **6**:101-124.

Melillo, J.M. 1996. Carbon and nitrogen interactions in the terrestrial biosphere: Anthropogenic effect. In B. Walker and W. Steffen (eds), *Global Change and Terrestrial Ecosystems*. Cambridge University Press, Cambridge: 431-450.

Melillo, J.M., A.D. McGuire, D.W. Kicklighter, B. Moore III, C.J. Vorosmarty, and A.L. Schloss. 1993. Global climate change and terrestrial net primary production. *Nature* **363**:234-40.

Metting, F.B. 1991. Biological surface features of semiarid lands and deserts. In J. Skujins (ed), *Semiarid Lands and Deserts. Soil Resource and Reclamation*. Marcel Dekker, New York: 257-293.

Metting, F.B. (ed). 1992. *Soil Microbial Ecology: Applications in Agriculture and Environmental Management*. Marcel Dekker, New York.

Metting, F.B. 1996. Biodiversity and application of microalgae. *J. Industrial Microbiol.* **17**:477-489.

Mummey, D.L., J.L. Smith, and G. Bluhm. 1999. Model estimation of nitrous oxide emissions from U.S. grasslands. *Environmental Management* (in press).

Oades, J.M. 1988. The retention of organic matter in soils. *Biogeochemistry* **5**:35-70.

Oechel, W.C., S.J. Hastings, M. Jenkins, G. Reichers, N. Grulke, and G. Vourlitis. 1993. Recent change of Arctic tundra ecosystems from a net carbon sink to a source. *Nature* **361**:520-526.

Parks, P.J. 1992. Converting marginal crop and pasture land to trees and forest. In R.N. Sampson and D.Hair (eds), *Forests and Global Change: Vol 1, Opportunities for increasing forest cover.* American Forests, Washington DC: 97-122.

Parton, W.J., D.S. Schimel, C.V. Cole and D.S. Ojima. 1987. Analysis of factors conrolling soil organic matter levels in Great Plains grasslands. *Soil Science Soc. Amer. J.* **51**: 1173-1179.

Paustian, K.H., G.I. Ågren, and E. Bosatta. 1997. Modeling the role of litter quality on decomposition and nutrient cycling. In G. Cadisch and K. E. Giller (eds), *Driven by Nature: Plant Litter Quality and Decomposition*. Wallingford, England: 313-335.

Paustian, K.H., E.T. Elliott, H.P. Collins, C.V. Cole, and E.A. Paul. 1995. Use of a network of long-term experiments for analysis of soil carbon dynamics and global change: The North American model. *Australian Journal of Experimental Agriculture* **35**:929-39.

Pierce, F.J., and E.J. Sadler (eds). 1997. *The State of Site-Specific Management for Agriculture*. American Society for Agronomy, Madison, WI.

Polglase, P.J., and Y.P. Wang. 1992. Potential CO_2-enhanced carbon storage by the terrestrial biosphere. *Australian Journal of Botany* **40**:641-56.

Post, W.M., W.R. Emanuel, P.J. Zinke, and A.G. Stangenberger. 1982. Soil carbon pools and world life zones. *Nature* **298**:156-159.

Post, W.M., R.C. Izaurralde, L.K. Mann, and N. Bliss. 1999. Monitoring and verifying soil organic carbon sequestration. In *Carbon Sequestration in Soils: Science, Monitoring and Beyond*, N.J. Rosenberg, R.C. Izaurralde, and E.L. Malone (eds). Battelle Press, Columbus, OH.

Potter, C.S., and S.A. Klooster. 1997. Global model estimates of carbon and nitrogen storage in litter and soil pools: Response to changes in vegetation quality and biomass allocation. *Tellus Series B Chemical and Physical Meteorology* **49**:1-17.

Potter, C.S., J.T. Randerson, C.B. Field, P.A. Matson, P.M. Vitousek, H.A. Mooney, and S.A. Klooster. 1993. Terrestrial ecosystem production: A process model based on global satellite and surface data. *Global Biogeochemical Cycles* **7**:811-41.

Ruimy, A., B. Saugier and G. Dedieu. 1994. Methodology for the estimation of terrestrial net primary production from remotely sensed data. *Journal of Geophysical Research* **99:**5263-5283.

Sage, R.F., and R.K. Monson. 1998. *C4 Plant Biology*. Harcourt Brace & Co., New York.

Savka, M.A., and S.F. Farrand. 1997. Modification of rhizobacteria by engineering bacterial utilization of a novel plant-produced resource. *Nature Biotechnology* **15**:363-368.

Scharpenseel, H.W., M. Schomaker, and A. Ayoub (eds). 1990. *Soils on a Warmer Earth.* Elsevier, Amsterdam.

Schiefelbein, J.W., J.D. Masucci, and H. Wang. 1997. Building a root: the control of patterning and morphogenesis during root development. *The Plant Cell* **9**:1089-1098.

Schimel, D.S., B.H. Braswell, E.A. Holland, R. McKeown, D.S. Ojima, T.H. Painter, W.J. Parton, and A.R. Townsend. 1994. Climatic, edaphic, and biotic controls over storage and turnover of carbon in soils. *Global Biogeochemical Cycles* **8**:279-293.

Schlesinger, W.H. 1985. The formation of caliche in soils of the Mojave Desert, California. *Geochemica et Cosmochemica Acta* **49**:57-66.

Schlesinger, W.H. 1992. Carbon storage in the caliche of arid soils: A case study from Arizona. *Soil Science* **133**:247-255.

Schlesinger, W.H. 1997. *Biogeochemistry. An Analysis of Global Change*. 2nd edition. Academic Press, San Diego.

Schlesinger, W.H., P.J. Fonteyn, and G.M. Marion. 1987. Soil moisture content and plant transpiration in the Chihuahuan desert of New Mexico. *Journal of Arid Environments* **12**:119- 126.

Sewalt, V.J.H., N. Weiting, J.W. Blount, H.G. Jung, S.A. Masoud, P.A. Howles, C. Lamb, and R.A. Dixon. 1997. Reduced lignin content and altered lignin composition in transgenic tobacco down-regulated in expression of L-phenylalanine ammonia-lyase or cinnamate 4-hydroxylase. *Plant Physiol.* **115**:41-50.

Siegenthaler, U., and J.L. Sarmiento. 1993. Atmospheric carbon dioxide and the ocean. *Nature* **365**:119-25.

Smith, J.L., and E.A. Paul. 1990. The significance of soil microbial biomass estimations. *Soil Biochemistry* **6**:357-396.

Stallard, R. F. 1998. Terrestrial sedimentation and the carbon cycle: Coupling weathering and erosion to carbon burial. *Global Biogeochemical Cycles* **12**:231-257.

Taylor, J.A., and J. Lloyd. 1992. Sources and sinks of atmospheric CO_2. *Australian Journal of Botany* **40**:407-418.

Tisdall, J.M., and J.M. Oades. 1982. Organic matter and water-stable aggregates in soils. *J. Soil Science* **33**:141-163.

Trexler, M. 1998. Forestry as a climate change mitigation option. Appendix K in Hertzog, H.J. (ed), *Proceedings of the Stakeholders' Workshop on Carbon Sequestration*. Massachusetts Institute of Technology, Energy Laboratory Report No. MIT-EL 98-002.

U.S. Department of Agriculture. 1990. *Agriculture Resources: Agricultural Land Values and Markets: Situation and Outlook Report.* AT-18, June 1990. Washington DC.

U.S. Department of Energy. 1999. *Carbon Sequestration, State of the Science: A working paper for roadmapping future carbon sequestration R&D.* U. S. Department of Energy, Office of Science and Office of Fossil Energy. Washington, DC.

U.S. Department of Energy. 1997. *Technology Opportunities to Reduce U.S. Greenhouse Gas Emissions*. Report by the National Laboratory Directors.

Vasievich, J.M., and R.J. Alig. 1996. Opportunities to increase forest growth and carbon storage on timberlands in the contiguous United States. In R.N. Sampson and D. Hair (eds), *Forests and Global Change: Vol 2, Forest management opportunities for mitigating carbon emissions.* American Forests, Washington D.C.

Walker, B.H., and W.L. Steffen (eds). 1996. *Global Change and Terrestrial Ecosystems*. Cambridge University Press, Cambridge, UK.

Watson, R.T., M.C. Zinyowera, and R.H. Moss (eds). 1996. *Climate change 1995: Impacts, adaptations and mitigation of climate change: Scientific-technical analyses*. Contribution of Working Group II to the Second Assessment Report of the Intergovernmental Panel on Climate Change. Cambridge University Press, Cambridge.

Weller, D.M., and L.S. Thomashow. 1994. Current challenges in introducing beneficial microorganisms into the rhizosphere. In F. O'Gara, D.N. Dowling, and B. Boesten (eds), *Molecular Ecology of Rhizosphere Microorganisms*. VCH mbH, Weinheim, Germany: 1-18.

Whetten, R.W., J.J. MacKay, and R.R. Sederoff. 1998. Recent advances in understanding lignin biosynthesis. *Ann. Rev. Plant Physiol. Plant Molecular Biol.* **49**:585-609.

Wollast, R., F.T. MacKenzie, and L. Chou (eds). 1993. *Interaction of C, N, P and S Biogeochemical Cycles and Global Change.* Springer Verlag, Berlin.

Wright L.L., R.L. Graham, A.F. Turhollow, and B.C. English. 1992. Growing short-rotation woody crops for energy production. In R.N Sampson and D.Hair (eds.*), Forests and Global Change: Vol 1, Opportunities for increasing forest cover.* American Forests, Washington DC: 123-156.

Young, E.C., and T.C. Osborn. 1990. Costs and benefits of the Conservation Reserve Program. *Journal of Soil and Water Conservation* **45:**370-373.

ACKNOWLEDGEMENTS

The authors thank Norman J. Rosenberg, R. Cesar Izaurralde, Rattan Lal, Steve Hildebrand, Greg Marland, Dave Shriner and Richard Stuckey for commissioning this paper and for organizing the workshop at which it was first presented. We also wish to thank William Schlesinger, Cynthia Cambardella, Norman Rosenberg and Elizabeth Malone for commenting on and reviewing the manuscript. We also thank Gary Jacobs (ORNL) for sharing information from ongoing planning activities sponsored by the Department of Energy.

COMMENTARIES ON "SCIENCE NEEDS AND NEW TECHNOLOGY FOR SOIL CARBON SEQUESTRATION"

The commentaries that follow have been edited slightly so as to highlight their key points. Although some or most of the commentators' suggestions were adopted by the white paper authors in their revisions, we have left them intact so the reader can gain an understanding of the discussions and dynamics of the process used at the workshop.

CYNTHIA A. CAMBARDELLA, *U.S. Department of Agriculture*

Human activities have profoundly altered the global C cycle in the past century, primarily by increasing atmospheric CO_2 as a result of land use changes and the burning of fossil fuels. In their discussion paper, Metting et al. have explored the scientific feasibility of C sequestration by terrestrial ecosystems as a possible mechanism to slow atmospheric CO_2 levels. The ability of soil to function as a sink for atmospheric CO_2 is specifically discussed. In this commentary, I will discuss several scientific problem areas that deserve research attention that were highlighted in the second and third sections of Metting et al.'s discussion paper.

Below-Ground C Fluxes

Although the role of soil and soil organic matter (SOM) in the global C cycle has been researched extensively for many years, critical gaps in our knowledge about the mechanisms responsible for cycling and partitioning of photosynthetically fixed C into the various forms of soil C still exist. Notably, there is a lack of information on below-ground C allocation and cycling.

Roots are important sources of organic matter to soil, yet the rates of C flux through root systems to soils remains largely unknown, despite a wide appreciation of its potential importance. These below-ground fluxes are important for maintaining soil C pools, stimulating soil biological activity, and enhancing the aggregate structure of soils. The principal reason for our lack of good quantitative information on below-ground C fluxes is methodological problems associated with estimating rates of root turnover from measurements of root biomass. For instance, simultaneous root growth, mortality, and decay can cancel one another out and lead to the observation that standing stocks of root biomass aren't changing with time. Our basic inability to accurately quantify rates of below-ground C flux limits our ability to quantify total plant productivity, which, in turn, limits our ability to quantify C inputs to soil and subsequently to evaluate soil C sequestration potential.

Biologically Active Soil Organic Matter and Soil C Sequestration Potential

Over the past couple of decades, considerable progress has been made towards understanding soil organic matter dynamics in temperate grassland systems. An important advance has been to integrate the structural and functional properties of SOM and to relate them both to soil biological processes. Evidence suggests that SOM turnover and consequently nutrient cycling and C storage potential are coupled to the formation and stabilization of soil aggregates.

The most labile form of biologically active organic C, conceptually defined as the active C pool, is rapidly cycled in soil and represents a relatively small proportion of the total soil organic C (3-5%). The importance of the active pool in maintaining soil fertility and controlling the supply of nutrients to the plant during the growing season has been extensively documented. However, the turnover time of the active pool is so rapid that its direct effect on soil C storage potential may be limited.

A significant proportion of biologically active organic C in soil is physically protected from decomposition through occlusion inside soil aggregates. Conceptually, this protected material comprises a pool of C that has an intermediate turnover time in soil on the order of 10-50 years and can represent up to 40% of the total soil C in native grassland systems. This intermediately labile pool of organic C is turning over at a timescale that is appropriate for assessing the impact of climate and land use change on ecosystem function, and therefore is the critical organic matter pool to monitor as a short-term relative indicator of longer term changes in soil C storage potential.

Models

The effect of climate change on global air temperature has been expressed as small, incremental increases over extended periods of time. Likewise, changes in total soil C that arise as a result of climate and land use change will only be verifiable experimentally over extended time periods because it takes long periods of time for SOM levels to reach a new steady state. In order to predict absolute (as opposed to relative) long-term soil C sequestration potential, we will need to resort to the use of mathematically based simulation models. However, a model must be generalizable in order to be a useful predictive tool, and rigorous validation and verification are required for model generalization. In addition, a simulation model is only as good as our understanding of the rates of material transfer and our ability accurately quantify the model's input and output variables.

N Fertilization Effects on Below-Ground C in Agroecosystems

Many aspects of agroecosystem management have the potential to influence SOM dynamics and the rate of C sequestration in soil. Nitrogen fertilization is one management component that can influence SOM in two major ways: (1) by altering the quantity and/or quality of plant C inputs to soil and (2) by influencing SOM mineralization rates and C loss from the soil. The effect of N fertilization on plant growth with respect to above-ground production is very well documented. However, N fertilization effects on below-ground production and the relative allocation of C to above- and below-ground components are poorly understood. The regulation of C cycling by N fertilization in agroecosystems has an enormous potential to influence SOM dynamics and C sequestration in soil.

Land Degradation in the Tropics

During the past two decades, increases in human population densities in the tropics and the increased demand for cultivated land have resulted in unprecedented rates of deforestation. Following the conversion of natural tropical forest ecosystems, the land is typically cultivated until productivity declines and the land is abandoned. Soil organic C stocks decline following conversion of native forests, primarily because biomass inputs are low, but also because the coarse fractions of SOM are mineralized during land clearing and subsequent cultivation. Therefore, the failure to convert tropical forest to sustainable managed systems is directly related to declines in SOM. In addition, cultivation nearly always results in extensive soil degradation and a loss of soil fertility, which reduces the capacity of the biota to sequester C.

Trees have the potential to effectively remediate degraded soils, in part because their root systems can tap into deeper nutrient sources that are unavailable to many cultivated plants. Tree crops are currently an important land use in the tropics and agroforestry has the potential to become an economically viable and environmentally sustainable form of land management in these areas. However, our ability to design sustainable tree-based systems is hindered by an insufficient understanding of the biological processes involved in the restoration of degraded soils in the tropics.

A Couple of Concerns

Perhaps we should reduce the emphasis currently placed on finding the "missing C" and increase our efforts at accurately quantifying global C budgets, especially soil C inventories, using standardized methods for sample collection and analysis. At the same time, an accurate accounting of current and historic land use, dominant soil types, and relative landscape position can be taken. If the sampling scheme is designed to consider the variability of soil C across space and time, we could use the measurements to verify existing simulation models.

The idea that we are "buying time" over the next 50 years or so while we develop new biotechnological solutions to the "climate change problem" results in a very visceral, knee-jerk reaction from this scientist. I tend to worry whenever I hear anyone assert that science will "fix" our problems with a new pill, or a new device, or a new hybrid. This notion that we as a species must control, manipulate, and dominate the natural world in order to insure our survival is truly enigmatic. It always begs the obvious question, "...how do you think we got ourselves into this mess to begin with...?"

NORMAN J. ROSENBERG, *Pacific Northwest National Laboratory*

This paper makes a strong case for advancing knowledge of the plant biochemical pathways, genetic and metabolic processes that determine how much carbon is sequestered in soils. It also raises many interesting questions about our knowledge (or lack thereof) of the underlying processes that control carbon sequestration in soils under current climatic conditions. The authors should also offer suggestions for research that can address the question of how these processes might be affected by changing climatic conditions. Climate change impacts might be direct—e.g., temperature and soil moisture effects on rates of soil respiration—or indirect via climatically driven land use change.

Of course, lacking knowledge, as we do, of the geographical distribution of any impending climatic change makes it impossible to forecast specifically where, when and how much C sequestration will be affected. But research can provide insights on the magnitude of potential changes. Similarly, it would be useful to know how the dynamics of soil inorganic carbon accumulation or loss might be affected by climatic change.

The authors note that recalcitrance of soil organic matter is associated with the soil clay component indicating that colloidal processes are important. Do we need to emphasize ways of increasing colloidal activity in soils or even increasing colloidal content, perhaps by means of soil amendments?

Overall, the authors have contributed useful and needed insights for a research agenda to address opportunities in soil carbon sequestration.

I agree strongly with the authors' comments on exploitation of underutilized native plant resources. Experiments with halophytic vegetation in coastal regions of California produce large quantities of biomass that can be harvested for energy or plowed into the soil to increase its carbon content. Even if the above-ground biomass is removed, the below ground carbon content increases as roots ramify more widely and penetrate more deeply.

WILLIAM SCHLESINGER, *Duke University*

Reacting to the potential to sequester carbon in soils, especially during the next few decades, those with greatest relevance to the Kyoto protocol, I am perhaps less optimistic than most. Human population will increase dramatically, requiring more land to be devoted to food production and more intensive food production than we have ever witnessed on Earth. Meanwhile, we are likely to lower the overall net primary productivity of the planet and channel an increasing fraction of the remaining productivity to our own needs, leaving less to enter the soil ecosystem for storage.

Attempts to increase the productivity of existing agricultural lands by application of nitrogen fertilizer carry hidden carbon "costs," in terms of CO_2 emissions to the atmosphere, and direct and dramatic impacts of excess nitrogen availability in our environment. Attempts to increase the productivity of marginal lands, largely by irrigation, often cause large releases of CO_2 to the atmosphere associated with the precipitation of pedogenic carbonate. All these ancillary costs must be subtracted from any putative potential to sequester atmospheric carbon dioxide in soils.

Metting et al. suggest that irrigating arid lands with groundwater will result in a sink for carbon in soils. Each mole of carbon sequestered is, however, balanced by a mole of CO_2 released to the atmosphere. Whether there is potential for a net sequestration of carbon by provision of excess Ca, as a raw material for carbonate deposition in arid soils, is unclear. If it derives from $CaCO_3$, there will be no sink for C in soils, only a redistribution. If it derives from igneous rock, we will need a thorough analysis of the carbon costs of providing this Ca to arid soils at a rate that is meaningful relative to the rise of atmospheric CO_2.

Metting et al.'s calculation (Table 1-4) of potential carbon sequestration in soils suggests that slightly over 6 times the pool of active carbon in most soils must enter that pool in any given year. I question whether this factor can possibly be so similar for forests ranging from boreal to tropical latitudes. Moreover, it is unclear how Metting et al. estimate potential net sequestration from a steady-state model.

Better crop and forest management can be implemented "now," but whether these will lead a significant net sequestration of carbon in soils is unproven (Table 1-5). To reduce uncertainty, I would recommend more intensive whole-system experiments, examining how different ecosystems, under different management regimes respond to rising CO_2 and added nitrogen. Free-Air CO_2 enrichment (FACE) technology allows these investigations in large areas, with and without human interventions. Carefully designed tillage experiments, with the objective of measuring carbon storage per unit of land area and soil depth, would be a welcome addition to a surprisingly meager literature.

Metting et al. discuss a number of promising approaches based on the genetic engineering of crops and soil microbes. Mostly they suggest that these approaches may mature during the next 10 to 25 years to the point of enhancing the net sink for carbon in soils. Certainly, the advances of plant molecular biology have been dramatic, but that generic engineering will make such rapid progress is hope, not reality.

2

Monitoring and Verifying Soil Organic Carbon Sequestration

Wilfred M. Post, *Oak Ridge National Laboratory*

R. César Izaurralde, *Pacific Northwest National Laboratory*

Linda K. Mann, *Oak Ridge National Laboratory*

Norman Bliss, *EROS Data Center*

Soils store a significant amount of the carbon (C) involved in the modern global cycle of this element. Globally, soils contain approximately 1,500 Pg C (1 Pg = 1×10^{15} g). As a component of this cycle, soils can be either net sources or net sinks of atmospheric carbon dioxide (CO_2). During ecological succession, interactions between biota (e.g., autotrophs and heterotrophs) and environmental controls (e.g., temperature, moisture) lead to the accumulation of carbon as soil organic matter. Soil organic matter is approximately 58% carbon. Globally, land use changes and agricultural activities during the last 200 years have made soils act as net sources of atmospheric CO_2. However, evidence from long-term experiments suggests that C losses attributable to oxidation and erosion can be reversed with soil management practices that minimize soil disturbance and optimize plant yield through fertilization (Cole et al., 1996; Dick et al., 1998; Hendrix et al., 1998; Janzen et al., 1998; Peterson et al., 1998; Rasmussen et al., 1998). These experimental results are believed to apply to large regions and suggest that carbon sequestration is occurring as a result of establishment of perennial vegetation, increased adoption of conservation tillage methods (US Midwest; Prairie Provinces in Canada), efficient use of fertilizers, and increased use of high yielding crop varieties (Buyanovsky, 1998) (Figure 2-1). It is possible that improved land management can result in significant increase in the rate of carbon inputs into the soil. Because of the relatively long turnover time of some soil carbon fractions, this could result in sequestration of a sizable amount of carbon in soil for several decades. Management of soil carbon can be an important option in reducing CO_2 concentration in the atmosphere. At this time, however, we lack accurate estimates about the land areas involved and the rate of soil organic matter changes that might be occurring under improved management.

The level of carbon sequestration in soils eventually needs to be known at different scales of resolution: field, region, national, and global. Estimations of soil carbon stocks have been made at national (Kern et al., 1998, Tarnocai and Ballard, 1994) and global levels (Post et al., 1985; Eswaran et al. 1995; Batjes, 1996). While there are soil quality monitoring programs that measure, among other things, soil

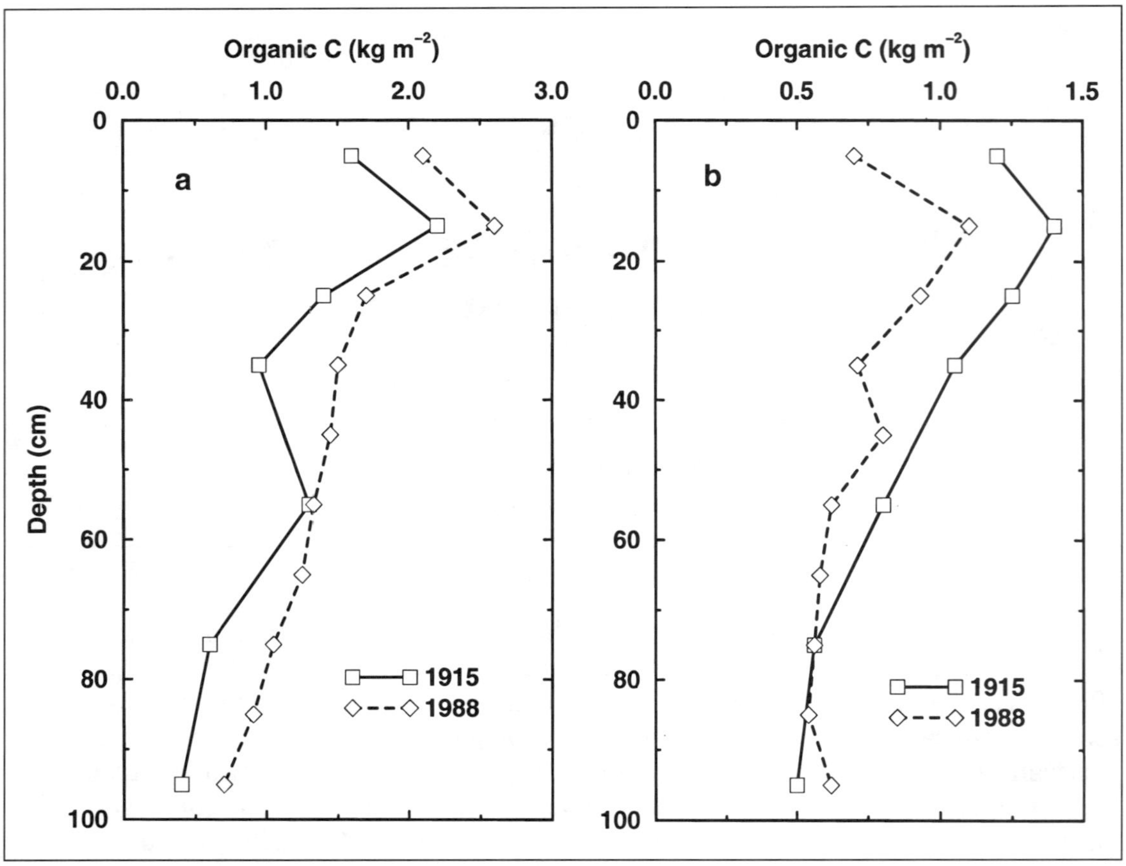

Figure 2-1. Soil organic carbon distribution with the soil profile for 1915 and 1988: (a) 3-year rotation (corn/wheat/clover) with 13.4 Mg ha^{-1} of manure and N; (b) with no treatment (note change in horizontal scale) (Modified from Buyanovsky and Wagner, 1998).

carbon changes, no official internationally agreed-upon method exists to verify or monitor changes in these stocks.

There is an urgent need to develop robust, science-based, flexible, and practical protocols for monitoring and verifying temporal changes in soil carbon. Here we discuss current methods for measuring carbon changes in soil, analyze sources of error that might arise from their application in monitoring and verification schemes, and outline practical, science-based procedures for a verification and monitoring plan. We also identify future research and data system development needs, including field sampling, laboratory measurements, simulation modeling, and remote sensing.

OVERVIEW OF SOIL PROPERTIES AND PROCESSES

The amount of carbon stored in soil is determined by the balance of two biotic processes—production of organic matter by terrestrial vegetation and decomposition of organic matter by soil organisms. Each of these processes is strongly controlled by physical and biological factors. The amount of organic matter produced depends largely on climatic conditions, soil water status, nutrient availability, and the growth and tissue allocation patterns of vegetation. In agriculture and silviculture, these physical and biological factors can be greatly influenced by management practices. The rate of decomposition depends on temperature and soil moisture status, chemical composition of organic matter, and soil physical structure. Although there may be a continuum of soil organic compounds in terms of their decomposability and turnover times, physical fractionation techniques are often used to define two relatively discrete soil organic carbon pools—light fraction and heavy fraction organic C (Christensen, 1996) (Figure 2-2). The light fraction is composed of free (not complexed with mineral matter) particulate plant and animal residues (Spycher et al., 1983). Part of the light fraction can be physically stabilized in soil aggregates through polycation bridging, organism glues, or organic-inorganic bonds, resulting in structures that entrap organic matter and protect it from decomposer organisms and their extracellular enzymes. Accumulation of light fraction carbon can be quite large in permanently vegetated soils (forests and grasslands). The turnover rate of light fraction carbon is linked to the dynamics of aggregate formation, and its quantity and turnover rate are greatly affected by cropping and tillage (Beare et al., 1994; Biederbeck et al., 1994).

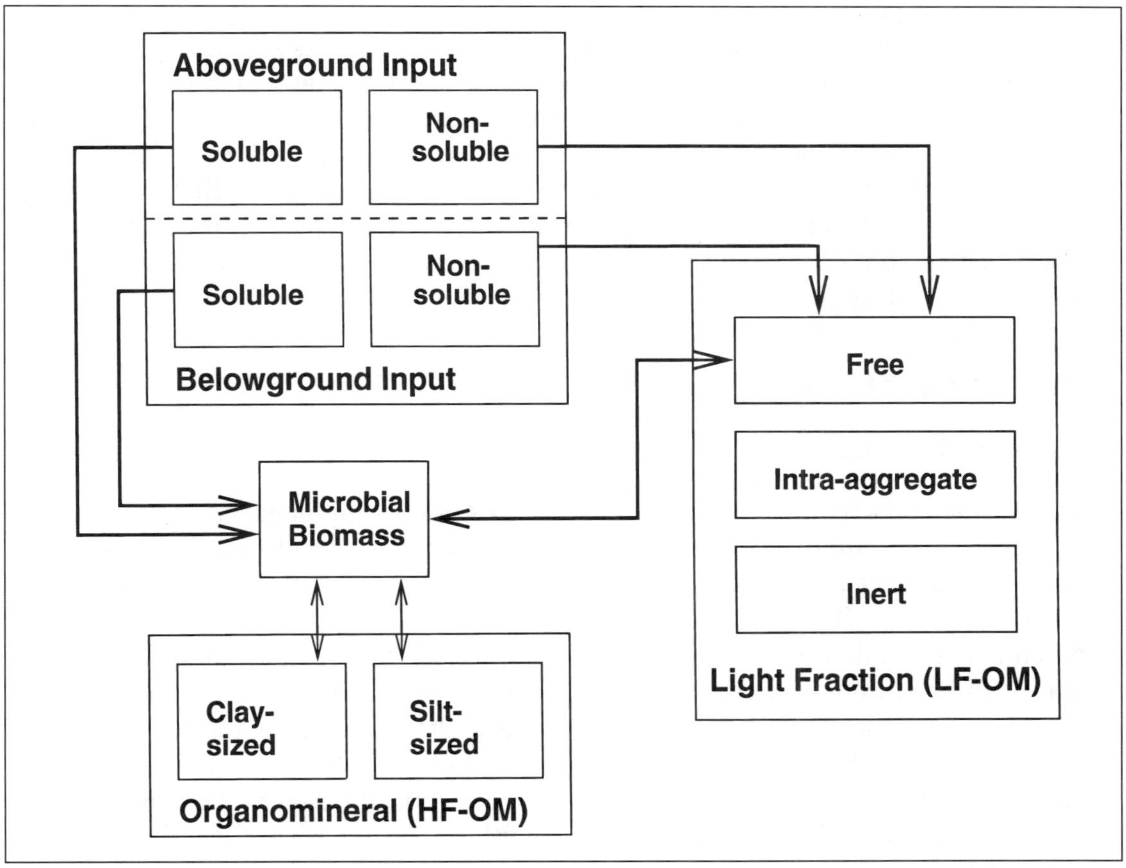

Figure 2-2. Compartment structure of soil organic matter model (modified from Christensen, 1996).

The heavy fraction of soil organic C has been further transformed by bacterial action and stabilized in organomineral complexes with clay and silt sized particles. This fraction constitutes the bulk of soil organic matter and is less susceptible to decomposition than the light fraction. Turnover times of the heavy fraction are on the order of decades; a portion that is very chemically recalcitrant has turnover times of 1,500 to 3,500 years (Parton et al., 1988; Jenkinson, 1990).

The decomposition rate of each soil organic matter fraction is usually well represented as a first order process, where the amount of decomposition per unit time depends on the amount of material subject to decomposition multiplied by the rate constant for the environmental conditions and type of material involved. The amount of material in each soil organic matter fraction at any time is a function of the previous history of additions relative to losses through decomposition and transformation. Therefore, the rate of change of soil organic matter content with change in management is dependent on the previous history of management and environmental conditions.

Measuring changes is complicated by the large amount of C that is relatively inactive on annual time scales. Changes in soil C on short time scales will be mostly in the relatively small light fraction, making it difficult to detect given the large amount of variability in the total amount of soil C introduced by different environmental conditions and histories, even for small areas. There are several approaches to dealing with these difficulties. We organize them into two categories: indirect methods which use information from previous studies combined with ancillary information, and direct methods which use measurements of C stocks or fluxes through time to determine changes in soil C amounts. The overall scientific enterprise depends on both the indirect and direct methods. The direct methods provide the field and laboratory measurements that form the foundation for the indirect methods. The indirect methods are required for extending the direct measurements in space and time, to give a realistic view of the variability in processes that influence soil organic carbon.

INDIRECT METHODS

Accounting Methods

To determine soil organic carbon pools and rates of change for large areas, it is necessary to extrapolate the relationships developed at the plot and field scale. Usually, the landscape is divided into patches, which are considered to have similar management and soil environmental conditions. The estimates for each unit area of patch are multiplied by the area of the total patch. The results for several patches can be summed to give an estimate for a region. The patches may be defined using soil survey, land cover, climate, and other spatial datasets. The relationships between the environmental factors and the soil carbon may be developed by measurements along chronosequences or from experimental results in agricultural, forest, or ecological research. (A chronosequence is a series of soil and vegetation units along a landform that can be used to discern the influence of time on pedogenic processes and ecological succession.)

If the patches become large, so that the assumptions of similar management and soil environmental conditions no longer apply, then the aggregated estimates become inaccurate. In some cases, the errors may tend to compensate, so that average conditions in a region might be used to estimate total conditions. For example, if soil organic carbon responds in a linear fashion to a factor such as precipitation, then using an average precipitation for the region in a model to predict the average soil carbon in the region might be successful. The low carbon values in areas of low precipitation would be compensated by the high carbon values in areas of high precipitation. Burke et al. (1990) found net primary production of short-grass prairie in Northeast Colorado to be linearly related to precipitation. To the extent that soil organic carbon contents respond linearly to carbon inputs, then averaging across a precipitation gradient may produce compensating errors.

Conversely, temperature gradients may be less reliable than precipitation gradients for estimating soil organic carbon. For a biologically active range of temperatures, the decomposition rate may vary exponentially with changes in temperature (a Q_{10} effect). If this is the case, then significant errors could be introduced by attempting to average across the gradient. It becomes necessary to subdivide the regions into smaller patches, or to explicitly model the known gradients that affect the soil organic carbon distribution and rates of change.

Another type of aggregation error may occur if soil texture varies over the region. Clay and silt-sized soil particles play important roles in physically protecting labile organic matter from decomposition. Identical management practices in closely located fields can have qualitative (Christensen and Johnson, 1997) as well as quantitative (Paustian et al., 1997) differences in soil organic carbon content related to variations in sand, silt, and clay content.

Stratified Accounting—the Geographical Information/Remote Sensing Approach

Inappropriate spatial averaging can be minimized by a systematic stratification of environmental factors that affect soil organic matter dynamics. In the United States, some very useful resources are available for constructing spatially detailed maps of soil organic carbon content and most of the important environmental factors that relate to soil organic carbon dynamics. Each of the spatial data sources has particular advantages.

Crop reporting statistics, forest inventories, and other information are gathered on a regular basis at a county or even finer spatial scale. This information is useful for estimating the amounts of organic matter produced each year. Various approximations must be made to estimate the quantities needed (kg C m^{-2}) from measurements actually collected (bushels of grain, tons of hay, wood volume increment, etc.).

Soil geographic databases stem from soil mapping efforts, such as the National Cooperative Soil Survey in the United States. These maps take into account the field experience of soil surveyors who integrate topography, microclimate, geology, and other soil forming factors, to describe the distributions of soils on the landscape. The data are typically focused on features of the landscape that essentially do not change during a mapping cycle (approximately 30 years). The State Soil Geographic (STATSGO) data base (NRCS, 1994) provides generalized soil data for the entire United States, and the Soil Survey Geographic (SSURGO) provides detailed soil information for selected counties (NRCS, 1995). Both these databases can be used to estimate soil organic carbon stocks using methods described by Bliss et al. (1995). A soil organic carbon map is being developed for North America (Canada, US, Mexico) (Lacelle et al., 1999). At a more generalized level, global soil organic carbon maps, based on FAO's Soil Map of the World, have been prepared by Sombroek et al. (1993) and Eswaran et al. (1995). As detailed as soil organic carbon maps may be for the United States, they can be improved by incorporating analyses derived from additional information such as digital elevation models, and land use and land cover data.

Digital elevation models (DEM), commonly having a horizontal spacing of elevation measures at every 30 m, are a rich source of data for modeling soil organic carbon on the landscape. The wetness of a soil is highly dependent on its landscape position, with wetter soils often occurring where the landscape is concave or in low-lying areas. Wet soils tend to accumulate more soil organic carbon. In some places, local geological features contribute to wet soils forming on uplands or side slopes. Digital elevation data provide a higher resolution of spatial patterns than standard soil maps, and a variety of derived datasets can be calculated, such as slope, aspect, flow direction, flow accumulation, stream length, and topographic position.

Databases of land use and land cover can be developed using remotely sensed data, in conjunction with elevation, ecoregion, and other data. Such databases can incorporate more information on how the landscape changes through time, with particular sensitivity to how vegetation changes with time. A project to develop global land use and land cover interpretations from satellite imagery has been completed at 1-km spatial resolution (Loveland et al, 1999). A related project is scheduled for completion in 1999 to map the land cover of the United States with a 30-m spatial resolution (Vogelmann et al., 1998).

Knowledge of the history of land use change is important for reconstructing past changes of soil organic carbon. One measure, the amount of land farmed, is available by county from the U.S. census since 1850. A dataset for land use history is important because past changes in cultivation, forest cutting and regrowth, and erosion are needed to understand present and potential future changes in C stocks and fluxes. In general, significant decreases in soil organic carbon have been associated with the conversion of natural vegetation to cultivation. Reconstructing the location and timing of these changes will lead to a better understanding of the current processes, and allow projection of the processes into the future.

Static maps of the current distribution of soil organic carbon stocks are good as starting points for assessing dynamic relationships (e.g., C sequestration potential). A simple starting point might be to assume that the potential for sequestration is greatest in those areas in which natural processes have resulted in the largest soil organic carbon accumulations. Humans can alter the processes, but the costs of doing so (in money, energy, and thus C emissions) may be large in areas where the natural processes do not effectively store C.

Availability of complete and detailed spatial and geographical information will increase the accuracy of using the stratified accounting approach to estimate soil organic carbon sequestration. Furthermore, sensitivity analyses of C measurements versus environmental factors should help reveal the extent to which the accuracy of spatial data layers is adequate or needs improvement. In this way, a strategy can be constructed for making regional and national estimates of soil organic carbon content and changes more precise by improving the most sensitive spatial data layers.

Future Remote Sensing Capabilities and Geographic Data

Remote sensing products soon to become available could prove very helpful for evaluating soil organic carbon, especially in regions that lack detailed geographical information. These include high resolution DEM and land cover characterization. A new global elevation dataset will be developed from the Jet Propulsion Laboratory's Shuttle Radar Topographic Mission, which is scheduled for launch in September 1999. The mission is projected to create 30- and 100-meter DEMs of the world between 60° N and 60° S latitudes within two years of the shuttle flight (http://www-radar.jpl.nasa.gov/srtm/). The MODerate-resolution Imaging Spectroradiometer (MODIS) sensor, to be on the Terra platform of NASA's Earth Science Enterprise, is scheduled for orbit in late 1999. This sensor may provide additional capabilities for spatial and spectral resolution that are not possible with the current Advanced Very High Resolution Radiometer sensor (AVHRR), the basis for the present global land cover datasets. Research using the MODIS sensor will enable interpretations of the remotely sensed images in terms related to biophysical processes of C accumulation on the landscape (e.g., leaf area index (LAI), the fraction of photosynthetically active radiation (fPAR), and net primary productivity (NPP).

The U.S. Geological Survey (USGS) is currently investigating the hypothesis that erosion and sedimentation may contribute to C sequestration. Geographic information (high resolution DEM, land use, soil type in particular) is being combined with models to estimate erosion and sedimentation of soil and associated organic carbon at landscape scales. A database of reservoir sedimentation is being used to benchmark the extrapolations of limited field data to the Mississippi River basin. Soil erosion models (such as WEPP, USLE) and soil organic carbon models (such as Century) are being used in this study.

There is a suggestion to extend the United State's Natural Resource Conservation Service (NRCS) pedon database to include more sampling on important "benchmark" soils that are spatially extensive and important for their C dynamics. Determinations of the amount and decomposition rates of various organic matter fractions (i.e., light fraction and heavy fraction) with distinct dynamical properties should also be included in future analysis (Paul et al., 1997).

Improvements in remote sensing and geographical information will enable better modeling of the regional distribution of soil organic carbon sources and sinks.

Process Models

Soil organic matter and ecosystem models will play an ever more important role in understanding land management and soil organic carbon sequestration relationships. Such models have been used to project changes in soil organic carbon content through time (Parton et. al., 1995) (Figure 2-3). Heterogeneity attributable to variation in initial conditions must be explicitly considered when using process models. Because the amount present and therefore the rates of change in each C fraction depend on management history, the histories must be accurately represented if the model estimates of soil organic carbon dynamics are to be realistic. When management history is well known for a period of at least 20 to 50 years, many soil organic matter models do well in simulating management-induced soil organic carbon changes (Smith et al., 1997). Generally, information on previous management history is less complete than needed to establish adequate initial conditions for models.

Many examples in the literature describe tests of soil organic matter models against short-term or long-term datasets at the plot or field scale. Ten of these were described and compared recently (McGill, 1996). There are fewer examples, however, of work comparing soil organic matter models and applying them at regional scales. Combining geographic information systems and remotely sensed data with soil organic matter models offers useful tools for understanding

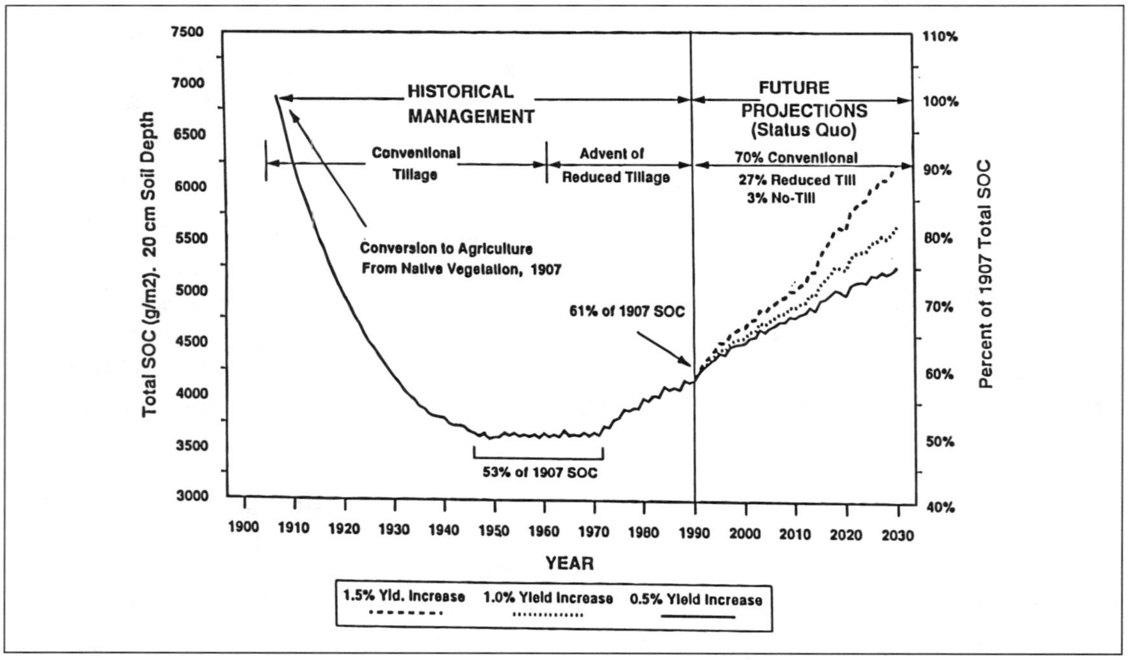

Figure 2-3. Century model-simulated soil organic carbon for the central US under the status quo scenario for three alternative levels of future crop yield increases (Parton et al., 1998; with permission).

of spatial and temporal soil organic carbon dynamics (Paustian et al., 1997; Burke et al., 1994; Falloon et al., 1998). In Canada, Izaurralde et al. (1999) compared six simulation models in order to select one that best fit data from long-term experiments and to scale up C storage at the regional level. Models specifically designed to simulate soil organic matter (RothC, CENTURY, DNDC, and SOCRATES) or that included soil organic matter dynamics as part of the soil-plant-atmosphere processes (ecosys, EPIC) were included in the comparison. Characteristics of four of the models are presented in Table 2-1.

The Canadian study selected the model SOCRATES for the aggregation experiments because it reproduced soil organic carbon dynamics best as determined in a series of long-term experiments in the Prairie Provinces and met both statistical (e.g. correlation, chi-square, lack of fit) and practical criteria (ease of use, simplicity and usefulness of input and output data). Three aggregation procedures were used to estimate carbon storage in the dominant soils of two ecodistricts having many soil-climate-management combinations. The first estimate (E1) was derived from simulating C storage in all the soils within each ecodistrict (14 for one and 7 for the second) under the dominant management regimes, including dairy, cattle, pigs, wheat, oilseeds, and grain. The second estimate (E2) simplified the modeling by hypothesizing that it was possible to simulate C storage using the dominant soil of each ecodistrict and by applying all the management systems characteristic of each ecodistrict. The third approach (E3) simplified even more

Table 2-1. Characteristics of four process-based models that simulate soil organic carbon dynamics: Century, DNDC, EPIC, RothC.

Century	DNDC	EPIC 5125	RothC	
Spatial scale	Plot to field	Plot	Plot to catchment	Plot to field
Time-step	Week	Hour and day	Day	Month
Time period for soil organic matter simulation	Months to centuries	Days to centuries	Years to centuries	Years to centuries
Soil profile	Uniform – 20 cm depth	Surface layers - 50 cm depth	Soil layers – up to 10	Uniform – 23 cm
Soil organic carbon pools	Three	Three	Two	Three
Texture effect	Yes	Yes	No	Yes
Weather data	Precipitation & temperature	Precipitation & temperature	Radiation, precipitation, temperature	Precipitation, temperature, open pan evaporation
Plant growth simulation	Cereals, grasses, trees	Cereals, grasses	Cereals, grasses, trees	No
Rotations	Multiple crops per year, unlimited rotations	Multiple crops per, year, rotations limited to 20 y	Multiple crops per year, unlimited rotations	None
Nutrients & fertilizers	N, P, S	N, P	N, P	None
Tillage	Many options	Many options	Many options	None
Erosion	No	No	Wind and water	No
CO_2 effect	Yes	Yes	Yes	No

the estimation of storage by assuming a dominant soil in each ecodistrict and one management that represented the major crops grown within each district. While in one ecodistrict estimates of annual soil organic carbon storage from the three aggregation methods were similar (35 Gg C y^{-1} for E1, 31 Gg C y^{-1} for E2, and 29 Gg C y^{-1} for E3, for an area of 650,000 ha), there were large discrepancies among the estimates made for the second ecodistrict (1 Gg C y^{-1} for E1, -4 Gg C y^{-1} for E2, and -12 Gg C y^{-1} for E3 over an area of 750,000 ha). Each aggregation method was evaluated based on accuracy, sensitivity to non-normal distributions of soil-climate management combinations, and effort involved.

These soil organic carbon models describe the processes of organic matter transformation, protection, and mineralization with different levels of detail and with different dependencies on environmental conditions. Consequently, each model has strengths and weaknesses for application under particular circumstances. These circumstances vary not only with physical and biological conditions of the region under study but also the amount of experimental experience and richness of climate, land use and geographical information available for the analysis. Understanding of the areal extent of the soil-climate-management combinations and their impact on model output is essential to decide on aggregation methods.

DIRECT METHODS

Field and Laboratory Measurements

The simplest way to report soil organic carbon content is as concentration—i.e., mass of C per unit mass of soil (g kg^{-1}). Frequently, concentrations of soil organic carbon are expressed as soil organic matter (soil organic matter = 1.724 x soil organic carbon). In C balance calculations, C concentration must be reported as soil organic carbon on an area (kg m^{-2}) or volume basis (kg m^{-3}). The calculation is not difficult but requires awareness about the vertical and horizontal variability of soil properties in order to avoid systematic errors. Converting estimates of C from concentration to mass requires data or estimates on soil bulk density, rock content, depth of sampling, and root components. Each of these physical characteristics has its own suite of associated uncertainties.

Uncertainties associated with bulk density estimates result from (1) methods of measurement and estimation; (2) temporal uncertainties from eroded (missing) surface soil, compaction and mixing of horizons through land use practices (livestock use, farm and forestry equipment); and (3) spatial heterogeneity (Kral and Hawkins, 1994). Methods for estimating bulk density are based on empirical relationships or on useful physically defined approximations such as the Adams (1973) equation:

$$BD = \frac{100}{\frac{\%\ SOM}{0244} + \frac{100 - \%\ SOM}{M\ BD}}$$

where 0.244 is the density of organic matter and *MBD* is the mineral bulk density. Mineral particle density is usually assumed to be the specific gravity of quartz (2.65 Mg m^{-3}) but can vary from 2.3 to over 3.3 (Culley, 1993). Actual mineral bulk densities are considerably less than rock bulk densities since soil consists of irregularly shaped mineral particles that allow large voids between them. Mann (1986) used measured bulk densities and soil organic carbon from more than 200 samples to calculate field values of MBD for the Adams (1973) equation. MBD values estimated in this way ranged from 0.7 to 1.7 Mg m^{-3}. Errors associated with measuring or estimating bulk density are not well known. Calculating the quantity of carbon depends on multiplying bulk density and soil organic carbon concentration, so accurate bulk density values are needed for monitoring changes in soil organic carbon content.

Rock or coarse fragment (> 2 mm) content must be estimated and subtracted from the soil volume prior to estimating total C on an areal basis. Care must be exercised in estimating rock content, especially in non-agricultural soils, which are often more rocky and less homogeneous. Collecting samples in rocky areas and obtaining adequate numbers of samples is difficult (Fernandez et al., 1993; Kral and Hawkins, 1994).

The characteristics of a column of soil can change over time. Seasonally, moisture content, crop roots, etc. will affect bulk density estimates. Methods that minimize such problems should be employed, such as obtaining bulk density samples without compaction and taking samples at the same time of year. Within the soil column, bulk density may increase over time because of compaction or decrease because of bioturbation, cultivation, or changes in structure resulting from changes in organic C, soil structure, and roots (Oades, 1993). To compare the carbon in two soils, the same mass of soil should be compared, rather than the same volume (Ellert and Bettany, 1995). If the carbon content of a soil has increased over time, it will likely have a lower bulk density, so a greater depth of soil will need to be sampled than in a comparable soil with no change in bulk density. In this manner, C storage of soils under different management or sampled at different times can be compared directly.

The lateral movement of soil by erosion and deposition can confound efforts to monitor changes in soil organic carbon storage. The estimation of C transported through erosion is an important component of annual C budgets. In research environments, erosion processes (i.e., detachment, transport, and deposition) can be either measured or estimated (Lal, 1994). Measurements of erosion are long term, involved and costly and, therefore, are likely to remain in the domain of research. Simulation models that deal with erosion, such as RUSLE (Revised Universal Soil Loss Equation), EPIC (Erosion Productivity Impact Calculator), WEPP (Water Erosion Prediction Project), and WEPS (Wind Erosion Prediction System), may be the more practical and appropriate tools to estimate erosion effects in operational projects of C sequestration. Methods based on the use of the isotope ^{137}Cs have been used to estimate total erosion or soil transport over long periods. However, it is questionable whether this method can yield the appropriate information over short periods of time (e.g., 5-10 years). Most likely, erosion estimates by simulation models, together with information on C content in sediments and sediment enrichment ratios, will be the most practical way to estimate C changes by erosion and deposition processes.

Root content adds another source of variability to estimates of C content. Coarse roots and rock fragments are removed in standard soil sample preparation by passing the sample through a 2-mm sieve (USDA-NRCS, 1996), but material from fine roots and dead root fragments is often retained. These fragments < 2 mm need to be handled in a consistent way—either included or excluded. The spatial variability resulting from the contribution of roots and litter to total SOM is magnified in agricultural lands, where spatial arrangement and spacing of row crops and fertilizer applications affect the distribution of SOM. Many grasses form dense clumps that present sampling difficulties in addition to affecting spatial heterogeneity in SOM.

Methods for measuring soil properties have been extensively documented and many are used as standards (Carter, 1993; Culley, 1993; Page et al., 1982; Klute, 1986; Magdoff et al., 1996; USDA-NRCS, 1996). These methods are generally used to estimate total C (organic C + inorganic C) and organic C in bulk soil. Measurements of C in bulk soil include both C fractions (light fraction and heavy fraction). Although bulk soil determinations are useful for monitoring C changes, they may not be able to detect such changes over short periods (e.g., 3-5 years). There is evidence that much of the early change in soil organic carbon occurs in the light fraction-C, and so measurement of changes in this fraction may be useful (e.g., Janzen et al., 1992). Incubation is another commonly used technique to estimate labile pools or light fraction-C. Soil incubation consists of measuring CO_2 evolved from fresh samples of soil over time, usually about 2 months (Elliott et al., 1994). Microbial biomass can also be measured by microscopy (Schmidt and Paul, 1982), or by chloroform fumigation followed either by incubation (Parkinson and Paul, 1982) or by extraction with K_2SO_4 (Vance et al., 1987).

Traditionally, soil organic matter has been fractionated by sequential extractions in alkali and acid solutions (Schnitzer, 1982). Humic and fulvic acids are soluble in dilute base, leaving humin as residue. Humic acid is insoluble in dilute acid solution, thus fractionating the humic materials further. Other methods to identify soil organic carbon fractions include infrared and nuclear magnetic resonance spectroscopy, and gel chromatography (Henry and Harrison, 1996). However, there is not yet much experience with these methods.

Minimum detectable differences—spatial and temporal

Temporal and spatial variability in the amount of organic matter from roots, litter, and microbial biomass combined with variability in measurements of bulk density, coarse fragments, and comparable sample depths contribute to high variability in estimates of soil organic matter at all scales. Wilding and Drees (1983 [cited in Buol et al., 1997]) reported the spatial variability of soil properties with 95% confidence interval and accuracy limit of ±10% of the mean: pH required a minimum of 10 samples, texture required 10-25 samples, and at least 25 were needed to establish horizon thickness. They concluded that for the more variable soil properties, such as soil organic C, estimates of the mean within ±10% at the 95% confidence level are "unrealistic because of the large number of samples needed." Mollitor et al. (1980) also reported the need for a relatively large number (about 20 to 30) of 250 cm^3 samples for estimates of mean soil organic C within ±10% at the 95% confidence level in a study of floodplain forests. Garten and Wullschleger (in press) developed a concept of minimum detectable difference as part of their study of changes in soil organic carbon content following the planting of switchgrass. They estimated that more than 100 samples (2.54 cm cores) would be needed to detect a 2-3% change in soil organic C, but that a 10-15% change could be detected with reasonable sample sizes (n=16) (Figure 2-4). They concluded that more variance estimates were needed, but that once the level of variation in carbon content is determined, the number of samples needed to detect a 10%, 25%, or other magnitude of change can be established for a given level of statistical power.

The heterogeneity of soil organic carbon concentration and its dynamic nature make it infeasible to obtain estimates of soil organic carbon change on annual or finer time scales from direct measurement. Monthly and single year changes are small, even with massive changes in soil organic matter input rates. Interannual variation in environmental conditions also complicates obtaining a clear picture of trends at all spatial scales. However, after sufficient time (e.g., 5 to 10 years) statistically significant differences in soil organic carbon have been observed in natural, inadvertent, and planned experiments (e.g., Izaurralde et al., 1998b; Post and Kwon, in press; Smith et al., 1996; Paul et al., 1997). More field-level methodological efforts are needed to ensure that existing differences can be consistently detected with an acceptable level of precision under various field conditions.

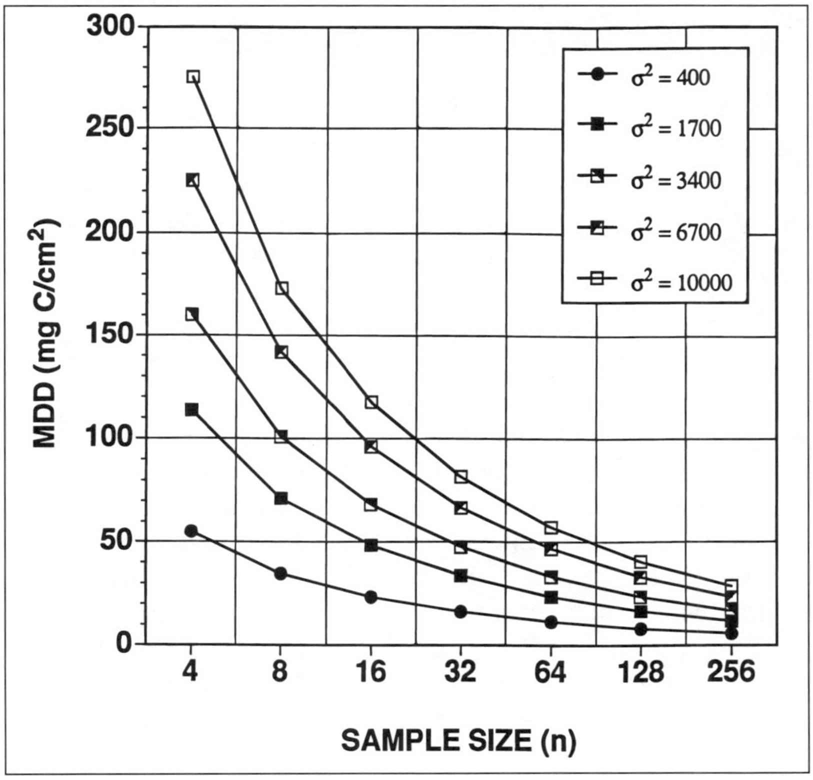

Figure 2-4. Calculated minimum detectable difference (MDD) in soil organic carbon inventory as a function of variance (σ^2) and sample size (n). The MDD is the smallest difference that can be detected ($\alpha = 0.05$) between two mean soil organic carbon inventories with 90% confidence ($1 - \beta$) given the average variance (mean square error from ANOVA) and the sample size (from Garten and Wullschleger (in press), with permission).

A role for eddy covariance for ecosystem CO_2 flux monitoring?

The vertical component of air movements over a vegetated surface (eddies) can be isolated and quantitatively measured as can CO_2 concentration associated with each eddy (Businger, 1986; Baldocchi et al., 1988; Verma, 1990). By correlating eddy size and CO_2 concentration for each upward and downward moving eddy, scientists can calculate the net uptake or release of C by the vegetation, i.e., the net ecosystem exchange. The integration of fluxes carried out over long enough periods yields estimates of change in C. The accuracy and precision of this methodology is improving as more experience is gained.

The net ecosystem exchange can be considered to have two components: changes in the carbon stock of the vegetation and of the soil. Changes in vegetation C stocks are generally easier to measure directly (except, of course, roots) and therefore changes in soil organic carbon stocks are usually assigned the difference between net ecosystem exchange and vegetation changes. The eddy flux procedures lead to an understanding of the changes in carbon fluxes on time scales of less than one year, and therefore will be useful in conjunction with the direct sampling and measurements of soil organic carbon described in previous sections. Some possibilities of improving the partitioning of net ecosystem exchange fluxes between plants and soil by using $^{13}CO_2$, $^{14}CO_2$, $CO^{18}O$ measurements are currently being explored. Coupling eddy covariance flux measurements with appropriate and thoroughly tested terrestrial biosphere models (e.g. SiB2, Sellers et al., 1996) could be an alternative method to identify portions of the net ecosystem exchange associated with soil organic carbon.

A "STRAWMAN" PLAN TO VERIFY AND MONITOR SOIL ORGANIC CARBON CHANGES AT REGIONAL SCALES

This strawman plan represents an outline of an initial monitoring plan for a region-wide determination of whether current land use practices are substantially changing the total amount of soil organic carbon. The steps involve selecting landscape study units, developing the measurement protocols, using remote sensing information and simulation models, and scaling results to represent the entire region. We borrow much of this plan from the GEMCo experience (Greenhouse [Gas] Emissions Management Consortium) (A. Donnelly, personal communication). Anticipating national and international efforts to reduce net CO_2 emissions, a group of energy industries in Canada formed the GEMCo. They have recognized the potential of soil carbon sequestration as a possible offset mechanism, together with the need to develop measuring and verification protocols (Ellert et al., 1999; Izaurralde et al., 1998a,b). Consequently, Canadian soil conservation associations have initiated work with energy industries and government agencies aimed at documenting soil carbon sequestration at the field level, with the hope of developing verifiable mechanisms for trading soil carbon credits. There are other efforts with similar objectives underway, notably a World Bank-Global Environmental Fund sustainable development project in Mexico which includes the assessment of soil carbon changes at a watershed level (C. Pieri, personal communication, December 3, 1998).

Identification of responsive sub-regions

The first step in developing methods to verify and monitor soil organic carbon changes at a given scale is to identify the sub-regions most responsive to changes in C management (e.g., counties within states or ecoregions within counties). Availability of information such as soil organic carbon maps, regional management history, and research data (e.g., Conservation Reserve Program (CRP), long-term experiments, ecosystem studies of wetlands, forest sites) would be useful during this planning step. Remotely sensed data and regional databases (soil survey, DEM, land use and land cover, census) would enhance the identification of areas with the greatest potential for C sequestration. Consultation with regional agronomists, soil scientists, field personnel and farmers' groups will help with the selection of pilot study areas and those to which the study results can be extrapolated.

A regional soil C baseline study conducted using past management history and soil databases would be a valuable starting point. Knowledge of whether or not soils are at steady state under a given management would allow initialization of simulation models in order to anticipate the effects of management alternatives. Such simulations cannot provide definitive answers, but should help narrow the management opportunities to be proposed to land managers of a given region. Management alternatives should be analyzed by teams of land managers, soil experts, economists, and industry representatives. These meetings would allow for the best selection of pilot study areas and those to which soil organic carbon changes can be extrapolated.

Practices that offer the greatest potential for carbon sequestration include conservation tillage, establishment of perennial vegetation, wetland restoration, extended rotations, nutrient management, and restoration of productivity on eroded land. Figure 2-5 depicts landscape components and operational tools from which to derive information in order to build a knowledge system for monitoring soil organic carbon.

Measurement of soil organic carbon changes

Changes in soil organic carbon can be determined by measuring the change in stocks or the change in fluxes. Sampling design must pay attention to horizontal and vertical heterogeneity, status of current soil organic carbon stocks in relation to carbon inputs and plant productivity, and movement of soil and associated soil organic carbon within and across the field. The following sections consider each of these factors.

Soil organic carbon sampling at field level. Once a field is selected to monitor soil organic carbon changes, the next step is to select the sampling area within the field. It is necessary first to investigate the spatial patterns present in a field and then to determine the best sampling methodology. The investigation of spatial patterns in fields should be done with the aid of soil survey information, aerial photographs, DEMs, and consultation with farmers and land managers. Knowledge of the patterns of spatial variation would most likely lead to a decision to stratify the

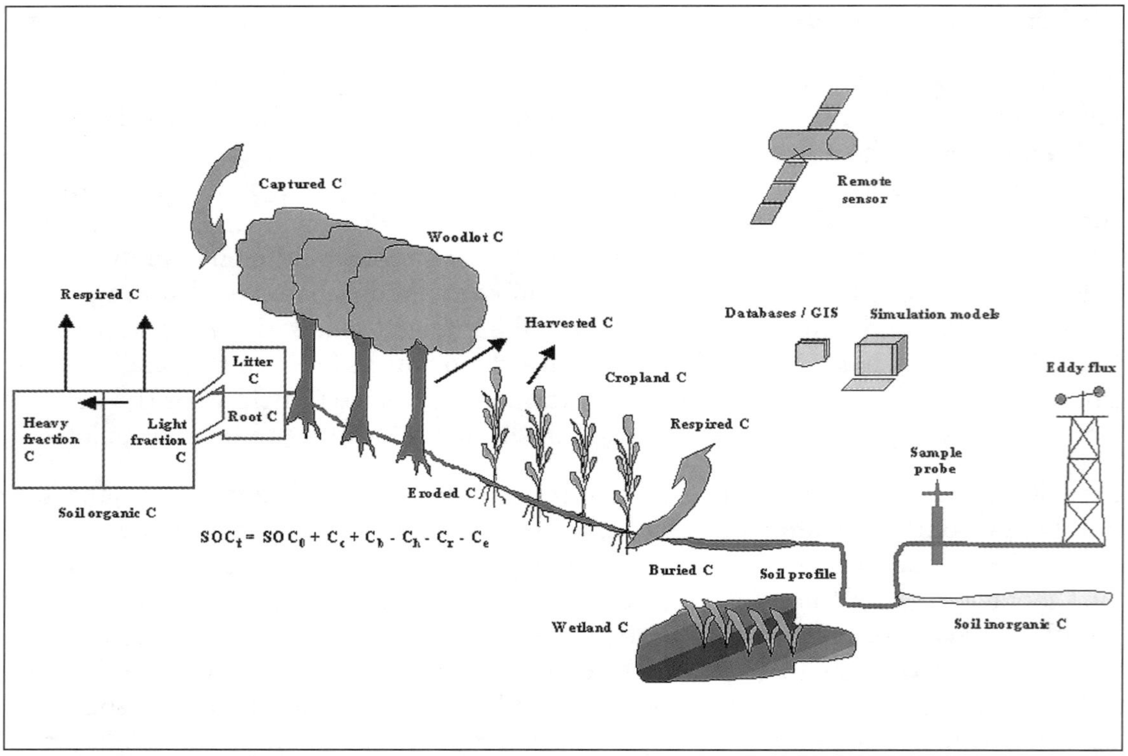

Figure 2-5. Carbon dynamics in a landscape and tools for its assessment. (Abbreviations: SOC = Soil Organic Carbon, 0 = initial time, t = time t, C_c = C captured by photosynthesis; C_b = C buried, Ch = C harvested (removed), C_r = C respired, C_e = C lost by erosion).

sampling by landscape position (e.g., knolls, back slopes, foot slopes). The next step would be to locate pilot areas to be sampled over time. Ellert et al. (1999) have a detailed plot layout designed to reduce errors induced by spatial variability. Returning to the same location instead of random sampling is considered essential to reduce variability.

Our aim is to determine soil organic carbon changes within short periods (e.g., 5 years) in response to a change in management. The difference in soil organic carbon mass at year 5 will yield true C gains or losses due to imposed management only if the soil was approximately at steady state at the start (McGill et al., in preparation). If possible, additional samples should be taken on similar nearby land maintained under traditional management during the study period. Differences in C storage between the new and the traditional management should yield the C loss "avoided" under the new management, another form of reducing C emissions (McGill et al., in preparation). When possible, the sampling areas should be geo-referenced using Global Positioning Systems (GPS) or other methods (e.g., metal rod burials and metal detectors). It would be helpful to have an instrument that could be carried to the field to make fast and accurate integrated measurements of soil organic carbon with depth. No such instrumentation currently exists but there are hopes for its development. Lacking such instruments now, we must use more traditional devices such as soil bulk density samplers, soil cores (hand or truck-mounted samplers), and gamma ray probes. All such methods and devices have advantages and disadvantages. Methods must be available and practical.

A practical method will only involve sampling the zones within the soil profile where changes in C stocks are likely to occur (i.e., near the surface, within the region of highest biological activity). A depth of 30 to 40 cm is considered appropriate in most crop- and range-lands. Greater depths may be necessary under perennial vegetation, especially grasslands. Soils should be sampled in increments of 7.5 to 10 cm. Information on sampling volume, which is needed to calculate soil bulk density, can be derived from the volume of the sampling device or other methods (e.g., filling sample holes with water or sand). Determination of the volume of rock fragments should be included with the field activity.

Sample processing should follow as quickly as possible, especially if biological determinations are to be made. Coarse roots (> 2 mm) should be separated by washing and their mass quantified. Soil samples should be ground to pass sieves < 2 mm. Laboratory analyses should be as complete as possible but at a minimum should include soil reaction (pH), carbonates, total C by dry combustion or soil organic C by the Walkley-Black method. If possible, the light fraction C by flotation or the acid resistant C fraction should be determined. Results of C (and N) analyses should be expressed on a carbonate-free, equivalent-soil mass basis. A portion of the soil sample should be dried and stored, preferably in glass containers for future analyses. This would be especially valuable if the method of C determination changes between sample periods. Then these stored samples could be reanalyzed with the new methods for comparison.

Determining changes in stocks over short periods (e.g., 1-2 years) is difficult because the changes to be measured are small compared to the size of the C stock and sampling variability. Minimum periods of three years or longer are needed to ensure the detection of these changes with acceptable confidence (e.g., 85-90% level of confidence). Determination of CO_2 fluxes in and out of the soil-plant system with eddy covariance techniques can also be used to estimate the change in soil organic carbon stocks. This approach is less invasive and integrates over the spatial heterogeneity in small areas. In some situations, it may be relatively simple to determine the soil component of CO_2 fluxes (croplands, rangelands). Forests and other ecosystems that have large potential for change in biomass quantity offer additional challenges.

Above-ground carbon. The quantification of C in plant biomass is essential to develop annual C budgets. In the case of woody species, this quantification can be made by allometry, checking and adjusting the calibration equations when necessary by using destructive sampling. In agricultural fields, crop yields can be estimated at the field level by the farmer, or by sampling the study area. With regard to the estimation of C inputs from straw and roots, these could be derived with straw-to-grain and root-to-shoot relationships acquired from measurements or from the literature.

Total above-ground C can be estimated by combining estimates of plant biomass, coarse woody debris (forest sites), fine litter and crop residue. Crop residue mass could be estimated by converting measurements of percent residue cover into residue mass through established relationships. Plant biomass can then be converted to plant C. Above-ground and below-ground C flows should be monitored on an annual basis. As with many studies, the periodic monitoring of agronomic (e.g., weeds, pests, disease) and ecological characteristics (e.g., species composition) will be important to explain the results.

Carbon losses and transport through erosion. As discussed earlier, estimation of C transport and deposition will emerge from the output of models such as EPIC, RUSLE, WEPP, and WEPS. The best effort should be made when characterizing landscape features (e.g., slope gradient and length, area) for initializing the models. Efforts should be made to ensure the comparability among model outputs.

Remote sensing

Remote sensing would play an important role in a plan to verify and monitor field changes of soil organic carbon in at least three ways: (a) to monitor biomass yield or leaf area index; (b) to derive current soil organic carbon maps, alone (Chen et al., submitted) or together with neural network methods (Levine and Kimes, 1998); and (c) to initialize (or correct) simulation models with estimates of LAI or biomass. Certain current sensors or platforms offer options for different measurement types and scale (e.g., AVHRR, AVIRIS, Landsat, SPOT). At this time, however, it is not possible to estimate soil organic carbon directly from remotely sensed imagery without relying on ground-based information (e.g., soil organic carbon values).

Simulation of soil organic carbon changes

Process based models offer plausible opportunities for projecting soil organic carbon trajectories in time and space (Izaurralde et al., 1999; Paustian et al., 1997). Improvements in the models are increasing their reliability for projecting the effect of management changes over several time scales. The models may be used for estimating changes in soil organic carbon over short periods of time, which direct measurements do not reliably detect. Verification of the model projections and the carbon sequestered should be done every 5 years or so by direct measurement.

Availability of soil, climate and management databases is essential to drive simulation models of soil organic matter and plant growth. Although the use of databases for simulating C changes under site-specific conditions is important, perhaps the most appropriate use of these databases will be for scaling up work at the regional level. Validation and calibration of soil organic matter models against long-term datasets should be a continuous process. It is unlikely that one model alone would serve all purposes: thus the need for model inter-comparisons.

The use of simplified models such as metamodels (Lakshminarayan et al., 1997) or a "soil carbon equation" is foreseen as a possible way to anticipate the possible impact of management practices on soil organic carbon changes under specified soil-climate conditions. These simpler models should be calibrated or developed from soil organic matter models and field data, and would be very useful tools for planning carbon sequestration strategies and perhaps for allocating payments or credits for carbon sequestration.

Scaling

Developing scaling procedures to relate soil organic carbon changes on individual plots or fields to a regional or national accounting is a formidable challenge. Table 2-2 describes the applicability of various methods to assess soil organic carbon changes at different scales of resolution. There are many issues of scale to be resolved. The fundamental method for scaling consists of subdividing the landscape into relatively homogeneous patches, applying field measurements or model results for each subdivision, and computing the area-weighted totals. The data and measurement requirements of this method are great, especially for regions of the world that have less information than the United States.

There are many opportunities to improve each aspect of these methods. These include improved data bases, accounting for processes that are currently not modeled, and refinements to models. Higher resolution data bases, such as 30m resolution datasets for digital elevation models and land use and land cover, provide the opportunity for decreasing the size of the landscape patches that are considered to be homogeneous, and thus reducing the variability of the measured variables. This leads to increased precision in estimating the stocks and fluxes of carbon for each landscape patch. With higher spatial resolution, there are more combinations of environmental conditions, which can give guidance to field efforts to observe conditions that have been under-represented in models. Additional processes can be modeled, such as the influence on the carbon cycle of erosion and sedimentation or atmospheric deposition of nitrogen. These may occur at different spatial or temporal scales than the early generation models have considered. With new data sources, and increased understanding of processes, the models will evolve to give more accurate estimates of carbon stocks and fluxes. Better models can be used in operational programs for encouraging carbon sequestration, including those involving payments or credits.

Table 2-2. The relationship between methods to quantify or estimate soil organic carbon changes and spatial scales.

	Sampling	Flux	Databases	Simulation modeling	Remote sensing
Field	X	X	X	X	X
Region		X	X	X	X
Nation			X	X	X
Globe			X	X	X

SUMMARY AND RESEARCH NEEDS

Accurate science-based methods are available for monitoring and verifying changes in soil C. Additional international collaborations should lead to codifying a set of methods into protocols that can be used to account for carbon sequestration in public agreements or private trading contracts. The level of precision will vary with the purposes for which the measurements are applied. Examples of purposes include (a) determining compliance with local, regional, and national laws or treaties regulating CO_2 emissions; (b) joint implementation projects; and (c) carbon credits and offsets. Current methods are effective for evaluating soil organic carbon changes at relatively low precision (20 to 50% error) and at widely spaced time intervals (minimum 3 to 5 years) with levels of effort that are reasonably affordable. Since relatively small amounts of C sequestered in soils could significantly reduce the rate of increase of atmospheric CO_2 and effectively buy time until low-cost methods for reducing CO_2 emissions are available, there will be considerable pressure to increase the reliability and precision of monitoring soil organic carbon changes on even shorter time scales.

Two basic approaches for monitoring and verifying C changes in soil were discussed. The first is based on indirect estimation of soil C stock changes over a specified period. The second is based on direct measurements of soil C pools and or fluxes of CO_2 between the atmosphere and land surface. Although both approaches are built on solid scientific principles, there is an urgent need to improve their operational accuracy and to make them cost effective enough to encourage their extensive use. Verification methods will have to include, at least temporarily, estimates based on incomplete knowledge of some fluxes and/or some temporal interpolation of anticipated long-term trends derived from experiments that may not directly apply to the situation at hand. Clearly, monitoring and verifying will have to rely somewhat on methods that fill in information in time and space that cannot be readily observed. Table 2-3 summarizes our understanding and vision with respect to current and future technologies for measuring soil C at the field level and making spatial and temporal projections of soil C changes.

Scientists are very interested in participating in the development of soil C monitoring protocols. However, in order to contribute to the development of these monitoring mechanisms, they need to know the economic and policy rules under which these mechanisms are anticipated to operate. Because so much is at stake, a multi-sector, multi-discipline, and multi-national effort is required if we are to make monitoring and verification of C sequestration in soils a useful and widely used procedure.

Table 2-3. Current and future technologies for monitoring soil C. (RS=Remote Sensing, LULC=Land Use and Land Cover, SAR=Synthetic Aperature Radar)

Technology	Current (1999 – 2001)	Mid term (2002 – 2007)	Long term (2008 – 2020)
C measurements			
Soil	Reduce sampling errors, improve root estimates	Non-destructive field measurement (exp.)	Non-destructive field measurement (routine, low cost)
Eddy flux	60 sites world wide	Expand to characterize significant landcover types	Routine, part of automated stations (when weather stations satisfy "upwind fetch" requirements; S. Verma, personal communication) (low cost)
Remote sensing	Low resolution LULC, absorbed PAR, hyperspectral (exp.), SAR (exp.)	High resolution, satellite-based hyper spectral, SAR, models (exp.)	High resolution, hyper spectral, SAR, models (routine)
C modeling	Models linked to databases; model intercomparisons	Models driven by RS input (exp.)	Real time simulation of land processes driven by RS
C accounting	Databases, maps, census, models (exp.)	Databases, maps, census, models, new sensors (refinement)	Databases, maps, census, models, new sensors (operational)

REFERENCES

Adams, W.A. 1973. The effect of organic matter on the bulk and true densities of uncultivated podzolic soils. *J. Soil Sci.* 24:10-17.

Baldocchi, D.D., B.B. Hicks, and T.P. Meyers. 1988. Measuring biosphere-atmosphere exchanges of biologically related gases with micrometeorological methods. *Ecology* 69:1331-1340.

Batjes, N.H. 1996. Total carbon and nitrogen in the soils of the world. *European J. Soil Sci.* 47:151-163.

Beare, M.H., M.L. Cabrera, P.F. Hendrix, and D.C. Coleman 1994. Aggregate-protected and unprotected organic matter pools in conventional- and no-tillage soils. *Soil Sci. Soc. Am. J.* 58:787-795.

Biederbeck, V.O., H.H. Janzen, C.A. Campbell and R.P. Zentner 1994. Labile soil organic matter as influenced by cropping practices in an arid environment. *Soil Biol. Biochem.* 26:1647-1656.

Bliss, N.B., S.W. Waltman, and G.W. Petersen. 1995. Preparing a soil carbon inventory for the United States using geographic information systems. In R. Lal, J. Kimble, E. Levine, and B.A. Stewart (eds), *Adv. Soil Sci: Soil and Global Change*, CRC Press, Lewis Publishers, Boca Raton, FL: 275-295.

Buol, S.W., F.D. Hole, R.J. McCracken, and R.J. Southard. 1997. *Soil Genesis and Classification*, 4th edition. Iowa St. Univ. Press, Ames, IA.

Burke, I.C., D.S. Schimmel, C.M. Yonker, W.J. Parton, L.A. Joyce, and W.K. Lauenroth. 1990. Regional modeling of grassland biogeochemistry using GIS. *Landscape Ecol.* 4:45-54.

Burke, I.C., W.K. Lauenroth, W.J. Parton, and C.V. Cole. 1994. Interactions of landuse and ecosystem structure and function: a case study in the central Great Plains. In P.M. Groffman and G.E. Likens (eds), *Integrated Regional Models: Interactions Between Humans and Their Environment,* Chapman and Hall, New York: 79-95.

Businger, J.A. 1986. Evaluation of the accuracy with which dry deposition can be measured with current micrometeorological techniques. *J. Clim. Appl. Meteorol.* 25:1100-1124.

Buyanovsky, G.A., and G.H. Wagner 1998. Carbon cycling in cultivated land and its global significance. *Global Change Biol.* 4:131-142.

Carter, M.R. (Editor). 1993. *Soil sampling and methods of analysis*. Can. Soc. Soil Sci., Lewis Publishers, Ann Arbor, MI.

Chen F., D.E. Kissel, L.T. West, and W. Adkins. Mapping soil surface organic carbon at the field scale with remotely sensed imagery. *Soil Sci. Soc. Am. J.* (Submitted).

Christensen, B.T. 1996. Matching measurable soil organic matter fractions with conceptual pools in simulation models of carbon turnover: revision of model structure. In D.S. Powlson, P. Smith, and J.U. Smith (eds), *Evaluation of Soil Organic Matter Models*, NATO ASI Series 1, Vol 38. Springer-Verlag, Berlin: 143-160.

Christensen, B.T., and A.E. Johnston. 1997. Soil organic matter and soil quality lessons learned from long-term experiments at Askov and Rothamsted. In E.G. Gregorich and M.R. Carter (eds), *Soil quality for crop production and ecosystem health. Dev. Soil Sci.* 25:399-430.

Cole, V., C. Cerri, K. Minami, A. Mosier, N. Rosenberg, and D. Sauerbeck. 1996. Agricultural options for mitigation of greenhouse gas emissions. In R.T. Watson, M.C. Zinyowera, and R.H. Moss (eds), *Climate Change 1995: Impacts, Adaptations, and Mitigation of Climate Change: Scientific-Technical Analyses*. Contribution of Working Group II to the Second Assessment Report of the Intergovernmental Panel on Climate Change, Cambridge University Press, Cambridge and New York: 744-771.

Culley, J.L.B. 1993. Density and compressibility. In M.R. Carter (ed.), *Soil Sampling and Methods of Analysis*. Can. Soc. Soil Sci., Lewis Publishers, Ann Arbor, MI: 529-539.

Dick, W.A., R.L. Blevins, W.W. Frye, S.E. Peters, D.R. Christenson, F.J. Pierce, and M.L. Vitosh. 1998. Impacts of agricultural management practices on C sequestration in forest-derived soils of the eastern Corn Belt. *Soil Till. Res.* 47:235-244.

Ellert, B.H., and J.R. Bettany. 1995. Calculation of organic matter and nutrients stored in soils under contrasting management regimes. *Can. J. Soil Sci.* 75:529-538.

Ellert, B.H., H.H. Janzen, and B. McConkey. 1999. A method for measuring soil carbon change on the Canadian Prairies. In R. Lal, J. Kimble, and R. Follett (eds.) *An international workshop on assessment methods for soil C pools*, Ann Harbor Press (in review).

Elliott, E.T., I.C. Burke, C.A. Monz, S.D. Frey, K.H. Paustian, H.P. Collins, E.A. Paul, C.V. Cole, R.L. Blevins, W.W. Frye, D.J. Lyon, A.D. Halvorson, D.R. Huggins, R.F. Turco, and

M.V. Hickman. 1994. Terrestrial carbon pools in grassland and agricultural soils: preliminary data from the Corn Belt and Great Plains Regions. In J.W. Doran, D.C. Coleman, D.F. Bezdicek, and B.A. Stewart (eds.), *Defining soil quality for a sustainable environment*. SSSA Special Public. No. 35, SSSA-ASA, Madison, WI: 179-192.

Eswaran, H., E. Van den Berg, P. Reich, and J. Kimble. 1995. Global soil carbon resources. In R. Lal, J. Kimble, E. Levine, and B.A. Stewart (eds.) *Advances in Soil Science: Soils and Global Change*. Lewis Publishers, CRC Press. Boca Raton, FLA: 27-43.

Falloon, P.D., P. Smith, J.U. Smith, J. Szabó, K. Coleman, and S. Marshall. 1998. Regional estimates of carbon sequestration potential: linking the Rothamsted Carbon Model to GIS databases. *Biol. Fertil. Soils* 27:236-241.

Fernandez, I.J., L.E. Rustad, and G.B.Lawrence. 1993. Estimating total soil mass, nutrient content, and trace-metals in soils under a low elevation spruce-fir forest. *Can. J. Soil Sci.* 73:317-328.

Garten, C.T., and S.D. Wullschleger. Soil carbon inventories under a bioenergy crop (Panicum virgatum): Measurement limitations. *J. Environ. Qual.* (in press).

Hendrix, P.F., A.J. Franzluebbers, and D.V. McCracken. 1998. Management effects on C accumulation and loss in soils of the southern Appalachian Piedmont of Georgia. *Soil Till. Res.* 47:245-251.

Henry, C.L. and R.B. Harrison. 1996. Carbon fractions in compost and compost maturity tests. In F.R. Magdoff, M.A. Tabatabai, and E.A. Hanlon, Jr. (eds). *Soil Organic Matter: Analysis and Interpretation*, SSSA Special Public. No. 46: 51- 67.

Izaurralde, R.C., K.L. Hauren-Kozyra, D.C. Jans, W.B. McGill, R.F. Grant, J.C. Hiley, and H.H. Janzen. 1999. Soil C dynamics: measurement, simulation and site-to-region scale-up. In R. Lal, J. Kimble, and R. Follett (eds.) *An international workshop on assessment methods for soil C pools*. Ann Harbor Press (in review).

Izaurralde, R.C., W.B. McGill, A Bryden, S. Graham, M. Ward, and P. Dickey. 1998a. Scientific challenges in developing a plan to predict and verify carbon storage in Canadian Prairie soils. In R. Lal, J. Kimble, R. Follett and B.A. Stewart (eds), *Management of carbon sequestration in soil. Adv. Soil Sci.* CRC Press, Inc., Boca Raton, FL: 433-446.

Izaurralde, R.C., M. Nyborg, E.D. Solberg., H.H. Janzen, M.A. Arshad, S.S. Malhi, and M. Molina-Ayala. 1998b. Carbon storage in eroded soils after five years of reclamation techniques. In R. Lal, J. Kimble, R. Follett and B.A. Stewart (eds.), *Soil processes and the carbon cycle. Adv. Soil Sci.* CRC Press, Inc., Boca Raton, FL: 369-386.

Janzen, H.H., C.A. Campbell, S.A. Brandt, G.P. Lafond, and L. Townley-Smith. 1992. Light-fraction organic matter in soil from long-term rotations. *Soil Sci. Soc. Am. J.* 56:1799-1806.

Janzen, H.H., C.A. Campbell, R.C. Izaurralde, B.H. Ellert, N. Juma, W.B. McGill, and R.P. Zentner. 1998. Management effects on soil C storage on the Canadian prairies. *Soil Till. Res.* 47:181-195.

Jenkinson, D.S. 1990. The turnover of organic carbon and nitrogen in soil. *Philos. Trans. R. Soc. London Ser. B 329:361-368.*

Kern, J.S., D.P. Turner, and R.F. Dodson. 1998. Spatial patterns in soil organic carbon pool size in the northwestern United States. P. 29-44. In Lal, R., R. Follett and B.A. Stewart (eds.), *Soil processes and the carbon cycle.* Adv. Soil Sci., CRC Press, Inc., Boca Raton, Fla.

Klute, A. (ed.). 1986. *Methods of soil analysis.* Part 1, Agronomy Monograph No. 9, American Society of Agronomy, Madison, Wisconsin.

Kral, D.M., and S.L. Hawkins (eds.). 1994. Erosion and productivity of soils containing rock fragments. SSSA Special Public. No. 13. 103 pp.

Lacelle, B., S. Waltman, N. Bliss, and F. Orozco-Chavez. 1999. The methods used to create the North American soil organic carbon digital database. In R. Lal, J. Kimble, and R. Follett (eds.) *An international workshop on assessment methods for soil C pools.* Ann Harbor Press (in review).

Lakshminarayan, P.G., P. Gassman, A. Bouzaher, and R.C. Izaurralde. 1997. A metamodeling approach to evaluate agricultural policy impact on soil degradation in western Canada. *Can. J. Agric. Econ.* 44: 277-294.

Lal, R. (ed.). 1994. Soil erosion research methods, 2nd edition. Soil Water Conserv. *Soil organic carbon*, St. Lucie Press, Delray Beach, FLA. 340 pp.

Levine, E.R., and D. Kimes. 1998. Predicting soil carbon in Mollisols using neural networks. In R. Lal, J. Kimble, R. Follett and B.A. Stewart (eds.), *Soil processes and the carbon cycle. Adv. Soil Sci.* CRC Press, Inc., Boca Raton, FL: 473-484.

Loveland, T.R, B.C. Reed, J.F. Brown, D.O. Ohlen, Z. Zhu, L. Yang, and J.W. Merchant. 1999. Development of a global land cover characteristics database and IGBP DISCover from 1-km AVHRR Data. *Int. J. Remote Sens.* (in press). (http://edcwww.cr.usgs.gov/landdaac/glcc/tablambert_na.html)

Magdoff, F.R., M.A. Tabatabai, and E.A. Hanlon, Jr., (editors). 1996. Soil Organic Matter: Analysis and Interpretation. SSSA Special Public. No 46. 67 pp.

Mann, L.K. 1986. Changes in soil carbon after cultivation. *Soil Sci.* 142:279-288.

McGill, W.B. 1996. Review and classification of ten soil organic matter (SOM) models. In D.S. Powlson, P. Smith, and J.V. Smith (eds), *Evaluation of soil organic matter models using long-term datasets.* NATO ASI Series I, Vol. 38. Springer-Verlag, Heidelberg: 111-132.

McGill, W.B., Y. Feng, and R.C. Izaurralde. Soil organic matter dynamics: from past frustrations to future expectations. (in preparation).

Mollitor, A.V., A.L. Leaf, and L.A. Morris. 1980. Forest soil variability on northeastern flood plains. *Soil Sci. Soc. Am. J.* 44:617-620.

NRCS (Natural Resources Conservation Service). 1994. State soil geographic (STATSGO) data base - data use information: Ft. Worth, Texas, U.S. Department of Agriculture, Natural Resources Conservation Service, Miscellaneous publication number 1492 (revised edition) [variously paged].

NRCS (Natural Resources Conservation Service). 1995. Soil survey geographic (SSURGO) data base - data use information: Ft. Worth, Texas, U.S. Department of Agriculture, Natural Resources Conservation Service, Miscellaneous publication number 1527 [variously paged].

Oades, J.M. 1993. The role of biology in the formation, stabilization and degradation of soil structure. *Geoderma* 56:377-400.

Page, A.L., R.H. Miller, and D.R. Keeney (eds). 1982. *Methods of soil analysis, Part 2. Chemical and microbiological properties*. Agronomy Monogr. No. 9 (2nd Edition). ASA-SSSA, Madison, WI.

Parkinson, D. and E.A. Paul. 1982. Microbial biomass. In A.L. Page, R.H. Miller, and D.R. Keeney (eds), *Methods of Soil Analysis, Part 2. Chemical and Microbiological Properties*. Agronomy Monograph No. 9 (2nd Edition). ASA-SSSA, Madison, WI: 821-830.

Parton, W.J., J.W.B. Stewart, and C.V. Cole 1988. Dynamics of C, N, P and S in grassland soils: a model. *Biogeochemistry* 5:109-131.

Parton, W.J., D.S. Ojima, and D.S. Schimel. 1995. Models to evaluate soil organic matter storage and dynamics. In M.R. Carter and B.A. Stewart (eds.), *Structure and Organic Matter Storage in Agricultural Soils*. Adv. Soil Sci. CRC Press, Lewis Publishers, New York: 421-448.

Paul, E.A., K. Paustian, E.T. Elliot, and C.V. Cole (eds). 1997. *Soil Organic Matter in Temperate Agroecosystems: Long-Term Experiments in North America*. CRC Press, Boca Raton, FL.

Paustian, K., E. Levine, W.M. Post, and I.R. Ryzhova. 1997. The use of models to integrate information and understanding of soil C at the regional scale. *Geoderma* 79:227-260.

Peterson, G.A., A.D. Halvorson, J.L. Havlin, O.R. Jones, D.J. Lyon, and D.L. Tanaka. 1998. Reduced tillage and increasing cropping intensity in the Great Plains conserves soil C. *Soil Till. Res.* 47:207-218.

Post, W.M., J. Pastor, P.J. Zinke, and A.G. Stangenberger. 1985. Global patterns of soil nitrogen. *Nature* 317:613—616.

Post, W.M., and K.C. Kwon. Soil carbon sequestration and land-use change: processes and potential. *Global Change Biol.* (in press).

Rasmussen, P.E., S.L. Albrecht, R.W. and Smiley. 1998. Soil C and N changes under tillage and cropping systems in semi-arid Pacific Northwest agriculture. *Soil Till. Res.* 47:197-205.

Schmidt, E.L. and E.A. Paul. 1982. Microscope methods for soil microorganisms. In A.L. Page, R.H. Miller, and D.R. Keeney (eds). *Methods of Soil Analysis, Part 2. Chemical and Microbiological Properties*. Agronomy Monograph No. 9 (2nd Edition). ASA-SSSA, Madison, WI: 803-814.

Schnitzer, M. 1982. Organic matter characterization. In A.L. Page, R.H. Miller, and D.R. Keeney (eds.), *Methods of Soil Analysis, Part 2. Chemical and Microbiological Properties*. Agronomy Monograph No. 9 (2nd ed.). ASA-SSSA, Madison, WI: 81-594.

Sellers, P.J., D.A. Randall, G.J. Collatz, J.A. Berry, C.B. Field, D.A. Dazlich, G.D. Colello and L. Bounoua. 1996. A revised land-surface parameterization (SiB2) for GCMs. Part 1: Model formulation. *J. Clim.* 5:676-705.

Smith, P., D. Powlson, and M. Glendining 1996. Establishing a European GCTE soil organic matter network (SOMNET). In D.S. Powlson, P. Smith, and J.U. Smith (eds.),

Evaluation of Soil Organic Matter Models Using Existing Long-Term Datasets. NATO ASI Series, Vol 38, Springer-Verlag, Berlin: 81-97.

Smith, P., D.S. Powlson, J.U. Smith, and E.T. Elliott (eds.) 1997. Evaluation and Comparison of Soil Organic Matter models. *Geoderma* 81:1-225.

Sombroek, W.G., F.O. Nachtergaele, and A. Hebel. 1993. Amounts, dynamics and sequestration of carbon in tropical and subtropical soils. *Ambio* 22:417-426.

Spycher, G., P. Sollins, and S. Rose 1983. Carbon and nitrogen in the light fraction of a forest soil: vertical distribution and seasonal patterns. *Soil Sci.* 135:79-87.

Tarnocai, C. and M. Ballard. 1994. Organic carbon in Canadian soils. In R. Lal, J. Kimble, and E. Levine (eds), *Soil Processes and Greenhouse Effect.* USDA, Soil Conservation Service, National Soil Survey Center, Lincoln, NE: 31-45.

USDA-NRCS (United States Department of Agriculture, Natural Resources Conservation Service). 1996. Soil Survey Laboratory Methods Manual. Soil Survey Investigations Report No. 42, version 3.0, U.S. Government Printing Office, Washington, DC.

Vance, E.D., P.C. Brookes, and D.S. Jenkinson. 1987. An extraction method for measuring soil microbial biomass. *Soil Biol. Bichem.* 19:703-707.

Verma, S.B. 1990. Micrometeorological methods for measuring surface fluxes of mass and energy. In N.S. Goel and J.M. Norman (eds), Instrumentation for Studying Vegetation Canopies for Remote Sensing in Optical and Thermal Infared Regions. Spec. Issue *Remote Sens. Rev.* 5:99-115.

Vogelmann, J.E., T. Sohl, and S.M. Howard. 1998. Regional characterization of land cover using multiple sources of data. *Photogramm. Eng. Remote Sens.* 64:45-57.

Wilding, L.P., and L.R. Drees. 1983. Spatial variability and pedology. In L.P. Wilding, N.W. Smeck, and G.F. Hall (eds), *Pedogenesis and Soil Taxonomy. I. Concepts and interactions.* Elsevier: 83-116.

COMMENTARIES ON "MONITORING AND VERIFYING SOIL ORGANIC CARBON SEQUESTRATION"

The commentaries that follow have been edited slightly so as to highlight their key points. Although some or most of the commentators' suggestions were adopted by the white paper authors in their revisions, we have left them intact so the reader can gain an understanding of the discussions and dynamics of the process used at the workshop.

DARWIN ANDERSON, *University of Saskatchewan, Saskatoon*

I agree that this is a reasonable effort at a big task and was pleased to learn of the Maryland workshop to refine the paper.

A few comments:

Carbon stored in soil organic matter is a substantial part of the global C cycle, estimated at 1500 Pg, and about equal to the atmosphere and biota pools combined.

Bulk density (the absence of BD values in almost all soil survey databases of soil profiles) is probably the biggest bugbear in expressing organic C on a mass/unit area basis. Also, calculating BD from soil organic C can get one into a sort of circular calculation where one variable determines the second.

I recommend that all material <2 mm be considered soil organic matter—separating out the residues, roots, etc. is too messy and variable.

CON CAMPBELL, *Eastern Cereal and Oilseed Research Centre, Agriculture and Agri-Food Canada*

In this paper the authors outlined their objective by stressing: "There is an urgent need to develop robust, science-based, flexible, and practical protocols for monitoring and verifying temporal changes in soil carbon." Consequently, they "discuss current methods for measuring carbon changes in soil, analyze sources of error arising from their application in monitoring and verification schemes, and lay out the basis for a verification and monitoring plan." Because I am not an expert in certain aspects discussed (e.g., use of remote sensing or eddy covariance methods), I do not address these aspects of the paper here. I comment on some of the other aspects with which I am more familiar.

Post et al. have prepared a well thought out and well written paper. There are a few pertinent points that may have been overlooked, however, or that may require further consideration or clarification. My comments address 10 issues.

Factors Affecting C Change

Post et al. suggest that C sequestration is a function of C added in biotic residues and C lost by decomposition. I agree, but would suggest that we need to account for C lost by erosion in this relationship. For example, if soil is merely moved from point A to point B in the landscape, it is not lost by decomposition. Therefore, in a C balance sheet, the C lost by erosion should be estimated using a model such as EPIC. This C must be added to C measured in the soil at time t, before the C balance can be estimated by modeling the two processes outlined by the authors above.

The authors correctly stress that changes in soil C are so small and so slow that it will take years to record significant differences due to management. Soil quality attributes, such as light fraction organic matter (LFC), respond more quickly than soil organic C (SOC) to changes in management (Campbell et al. 1999a,b). It would be desirable though difficult to establish quantitative relationships between changes in LFC and SOC because of the dynamic nature of LFC fractions.

Measuring C Changes in Soil

The authors dealt thoroughly with some of the difficulties we face in measuring soil C changes. I shall comment on some of the points they raised and mention a few others.

Do we take random samples or do we take samples in a systematic manner at fixed predetermined points? There is no doubt that the latter provide more precise results, and this is the method recommended by Ellert et al. (1998) as mentioned by the authors. It is interesting to note the influence this systematic sampling may have on our results. In a study conducted in the 1950's at Swift Current, Saskatchewan, four soils were sampled periodically to assess changes in soil organic matter (SOM) after breaking sod. Ten replicate samples, stepped off from the same fixed point, were taken each year. This procedure yielded fairly precise results. However, the changes in total N (also C) calculated over the 12-year period were small compared to those obtained in a similar study using only 3 replicate samples taken every 3 years over 12 years. Use of the former method increased the precision of sampling and thus reduced the probability of making a Type I error (rejecting the null hypothesis when the null hypothesis is true).

To properly compare changes in C over time, we should convert C concentration to a mass basis, and preferably to a mass per equivalent depth basis (Ellert and Bettany, 1995). Until recently, most data in the literature have been expressed on a concentration (%) basis; and even when bulk density was used to convert concentrations to mass, this was done on a mass per volume basis. Other factors, such as sampling time and soil moisture content at the time of sampling, may increase variability in the data collected and make results more difficult to interpret.

To minimize the problems related to bulk density I believe the most accurate densities may be obtained if the soil can be sampled without compaction (e.g., digging down to a depth and going in laterally with a spade). However, this is a slow process and can quickly destroy the relatively small experimental plots we usually employ. A second approach may be to use a soil coring device but take samples at the same time of year each time. My preference is to take the soil samples after harvest because the soil is usually dry and less likely to compact despite its texture.

Sample Handling and Analysis

To determine SOC, samples do not have to be analyzed soon after they are taken as suggested by Post et al. They should be sieved, air-dried and stored (preferably in glass) until analysis, even years later.

I prefer to use combustion analysis of some sort and would definitely not use the Walkley Black technique for accurate measurement of C. I suggest determining total N as well as organic C because C/N ratios (outside the range of 9/1 - 11/1) can signal inaccurate measurements. It is wise to reanalyze soil samples previously analyzed, when analyzing your new samples, as a check on analytical accuracy from your laboratory.

The question of what to do about macro crop residues that are left on the sieve is an important one. Perhaps it is best to measure and record this material separately. This is particularly relevant when we compare no-tillage with conventional tillage. Even so, the core samplers we use are often too small in diameter to get a true representation of the residues on no-till land; therefore, we may be underestimating C in these treatments in our comparisons.

Another sampling problem similar to the one mentioned last is how to sample for C in grassland. If we use a soil corer, don't we miss some of the crown? Should part of the plant crowns be included in the sample (like we do with surface residues in no-till)? Where should we position the corer when sampling? It is not as straightforward as sampling conventionally tilled cropland where crop residues are mixed into soil.

Spatial Variability

To what extent is spatial variability a function of soil textural inconsistencies at a site? There is a direct relationship between C sequestration and clay content. Thus, if the texture of a site varies appreciably, SOC contents may not reflect the influence of treatments but rather the soil textural effect on SOC. We believe that this phenomenon helped us explain inconsistent results observed in rotation studies at Swift Current.

Another observation concerns the assumption that adjacent land of a similar soil type and vegetation has a similar starting C as an adjacent test plot. This assumption is commonly used when original measurements were not made or are not available. We should be aware that this assumption might be invalid at times and lead, therefore, to erroneous conclusions.

Process Models

I agree with Post et al. that process-based models are destined to play an important role in estimating and verifying C changes in soil. I also agree that a good knowledge of previous cultural and environmental history of a site in question is necessary so as to allow accurate initialization of the model. If the model is not properly initialized the accuracy of any extrapolation by the model is likely to be erroneous.

Defining the size of C pools and their rate constants is also critical. But these are not easy to attain, and they vary with soils and conditions. I am not confident that there are any process-based models currently available that can accurately estimate *short-term (<10 yr) C changes* in soils. The Century model may be effective enough when used for long-term simulations; but it is the short-term simulations that are of main interest to the solution of our present problems (Campbell et al., 1998b).

Post et al. correctly remind us that models need to be continually validated if their output is to be used accurately for interpolation or extrapolation. Unfortunately, I suspect, many users of models are not doing this. In fact, some are using the model results of other users as the basis for their own validation. Thus, if the users of earlier models have made errors, so will those following after.

Post et al. have painted an encouraging projection of how we may be able to use improved methods of mapping, topography, and ground cover, using GIS and DEM, to estimate C changes over landscape and time. However, in the long run, it seems to me that it is how well we can estimate (with models) the relationships between C inputs, weather, and C changes in soils, that will determine how accurately we will be able to extrapolate (or interpolate) C changes in space and time. Unfortunately, despite claims of modelers, the latter exercise seems wanting at the present time for short-term estimates, even on specific sites where their cultural history and weather is well known. In my view, this is where we need to concentrate our greatest efforts.

Use of Simplified Models

The authors suggest that there is a place for use of simplified models or soil C equations as a means of anticipating the impact of management practices on changes in SOC under specified conditions. I completely agree with this sentiment, and we have recently presented our approach at the 1998 American Society of Agronomy Meetings in Baltimore (Campbell et al. 1998a).

Which Soils Are Most Likely to Provide Increases in C?

The authors suggest that if we wish to develop methods to monitor and verify changes in SOC, then we will need to identify regions that will be most responsive to changes in C when management is changed. They go on to mention techniques (remote sensing, regional databases such as soil survey, land cover) that should help us identify such sites. Further, they suggest that assistance should be sought from knowledgeable soil scientists, farmers, and regional agronomists. However, Post et al. didn't really define what criteria will be used in deciding which areas will be most likely to sequester more C. From their earlier discussion, one factor, the current C content of the soil compared to that which is (or was) present under native conditions seems a good one. That is, the greater the difference between the latter two, the greater should be the potential to increase C. Other factors, such as soil texture and strength of the driving variables (since they control primary production and thus C inputs) should also be considered.

Practices that Have Potential to Promote C Sequestration

The authors listed a set of practices that have the potential to increase C sequestration. I agree with most of those listed, but I would like to make three observations on this matter.

1. It is not just extending a rotation (less fallow) that is important, it is also having a proper crop mix (e.g., inclusion of legumes). Monoculture is especially undesirable.

2. None of the practices listed will result in an increase in C unless the weather cooperates. After all, C sequestration depends on C inputs, which in turn depends on favorable weather.

3. There is a common belief that the conversion of cropland to grassland will invariably lead to an increase in soil C. Disregarding a reduction in erosion, is this always the case? I don't believe so. For example, in a crop rotation study initiated at Swift Current in 1987, we seeded one treatment to crested wheat grass. The study was set up on good Research Station land that had grown cereals previously. While all other wheat-based rotations were fertilized based on soil tests, the grass received N+P to replace that removed in hay each year. Surprisingly, we found that the grass treatment had the lowest soil C in 1993 and 1996 because it also had the lowest net primary productivity.

Depth of Sampling

Post et al. suggest that the depth of sampling should be 30-40 cm in 7.5-10 cm increments. I think these are reasonable choices in most instances. However, it might be wise to consider the rooting depth and distribution of roots for the crops being grown. Alfalfa and some grasses have a considerable amount of roots below the 40-cm depth. In the Canadian prairies –where shallow tillage prevails—we rarely find significant differences in SOC below 15-cm depth and sometimes none below 7.5 cm.

Another question on this topic is how do we decide what depths should be included in our C balance exercise? I use statistical significance, root distribution pattern, and depth of tillage to help me decide. Another consideration here is, if estimating the root mass from straw mass (straw/root ratio), when should roots and shoots be measured? Bear in mind that for annuals, roots reach a maximum at flowering, while shoots reach theirs at harvest. I use ratio of shoot at harvest to root at flowering, reasoning that some roots have already started decomposing in the soil (and sloughing) which I will not have measured when sampled at harvest.

Research Needs

There is a critical need to improve our soil C models so that they can at least come close to estimating correctly the direction of change in the short term (<10 yr.) in response to crop management and/or weather; at present they do not do this (Campbell et al., 1998a). We also need to determine why changes in C in response to an imposed management might vary depending on ecoregions. For example, why does no-tillage often appear to result in C increase in the prairies, while in Ontario and Quebec we often observe no change or C decrease? Could this be related to the fact that although more humid conditions in eastern Canada support greater crop production, they also increase the rate of soil C decomposition? (Campbell et al., 1998*c*). If we can solve the above problems satisfactorily, then I believe the rest will fall into place, at least in the developed countries of the world.

References

Campbell, C.A., E.A. Paul, and W.B. McGill. 1976. Effect of cultivation and cropping on the amounts and forms of soil N. Pp. 7-101 In *Proc. Western Canada Nitrogen Symposium.* Alberta Soil Science Workshop, Calgary, Alberta.

Campbell, C.A., B.G. McConkey, R.P. Zentner, F.B. Dyck, F. Selles, and D. Curtin. 1996a. Carbon sequestration in a Brown Chernozem as affected by tillage and rotation. *Can. J. Soil Sci.* 75:449-458.

Campbell, C.A., B.G. McConkey, R.P. Zentner, F. Selles, and D. Curtin. 1996b. Long-term effects of tillage and crop rotations on soil organic C and total N in a clay soil in southwestern Saskatchewan. *Can. J. Soil Sci.* 76:395-401.

Campbell, C.A., E.G. Gregorich, R.P. Zentner, G. Roloff, H. Janzen, K. Paustian, W. Smith, B-C Liang, and B. McConkey. 1998a. Carbon sequestration on the Canadian Prairies—Quantification of short-term dynamics. Presented in symposium on Soil Management for Carbon Sequestration to mitigate the Greenhouse Effect II at the ASA meeting, Baltimore, MD, October 18-22.

Campbell, C.A., F. Selles, G.P. Lafond, B.G. McConkey, and D. Hahn. 1998*b*. Effect of crop management on C and N in long-term crop rotations after adopting no-tillage management: Comparison of soil sampling strategies. *Can. J. Soil Sci.* 78:155-162.

Campbell, C.A., G.P. Lafond, R.P. Zentner, and T.L. Roberts. 1998c. No-till management requires proper fertilization. *Better Crops* 82(4):14-15.

Campbell, C.A., V.O. Bierderbeck, G. Wen, R.P. Zentner, J. Schoenau, and D. Hahn. 1999a. Seasonal trends in selected soil biochemical attributes: Effects of crop rotation in the semiarid prairie. *Can. J. Soil Sci.* (in press).

Campbell, C.A., G.P. Lafond, V.O. Biederbeck, G. Wen, J. Schoenau, and D. Hahn. 1999b. Seasonal trends in soil biochemical attributes: Effects of crop management on a Black Chernozem. *Can. J. Soil Sci.* (in press).

Ellert, B.H. and J.R. Bettany. 1995. Calculation of organic matter and nutrients stored in soils under contrasting management regimes. *Can. J. Soil Sci.* 75:529-538.

Ellert, B.H., H.H. Janzen, and B. McConkey. 1998. A model for measuring soil carbon change on the Canadian Prairies. P. 14 in *Abstracts of International Workshop on Assessment Methods for Soil C Pools*, November, Columbus, OH.

ARLAN FRICK, *University of Saskatchewan, Saskatoon*

This paper is comprehensive in its review of methodologies for monitoring and verifying soil organic carbon stocks. There were several ideas that I was not familiar with, and that have interesting and promising prospects. I have only a few specific comments as outlined below.

In the discussion of estimation of soil erosion for SOC removals, it could be also noted that in some landscapes soil is simply redistributed over short distances and there is no net loss or gain of soil, or SOC, from the system. Therefore, soil redistribution, and not just soil erosion, should be considered. In this situation accurate estimates of soil redistribution are still important because the loss or gain of soil and SOC affects other soil and ecosystem processes that in turn influence SOC dynamics. For example, SOC loss from an upper slope reduces fertility, etc., leading to lower productivity and carbon inputs. In a depositional area, SOC could be buried at greater depths, influencing its turnover rates, etc.

In the discussion of variability of SOC measurements, I assume the authors are discussing variability due to landscape position/topography, how much of the variability is due to this, and how much due to other factors.

In the discussion of long time (5 to 10 years) needed to measure significant differences in SOC changes, what are the implications of this for the 2008-2012 monitoring period proposed in the Kyoto Protocol?

With regard to using DEMs to sample soil organic carbon at the field level, the 30 or 100 m DEMs could be useful for regions for long, gentle slopes. However, where there is more complex topography, much variation that would be very important for SOC would not be seen in the 30 m, and especially the 100 m DEM.

The authors make a good point about accurate initialization of SOC pools needed for model runs.

Eventually, it would be useful to link erosion models with the SOC models (e.g., CENTURY). Thus, the output of the erosion model on SOC redistribution could be directly used in the SOC simulation, and the changes in SOC would be fed back into the erosion model. The interrelationships between erosion and SOC changes could be simulated this way.

DOUGLAS JOHNSON, V. HAARMANN, Y. LI, N. MANSPEIZER, A. MARZOUK, *Clark University*

Twenty-two years ago the United Nations organized a conference (UNCOD) on the climatic, ecological, economic, social, and technological dimensions of desertification. The purpose of the conference was to generate support for a global struggle against the degradation of the world's drylands. Rallying this support was predicated both on presenting the very best current scientific understanding of how desertification processes are set in motion and on identifying the strategies that should be followed to arrest and reverse dryland degradation. Integral to this purpose was the widely shared perspective that enough was known about the physical process of desertification to understand how degradation developed and spread and that sufficient anti-desertification technology and experience were in existence to combat desertification successfully. Only three ingredients, it was thought, were lacking: political will to tackle the problem; capital resources to fund anti-desertification efforts; and the cooperation of misguided traditional land managers whose arcane practices were at the root of desertification's spread and whose commitment to tradition slowed the spread of new technology. The current workshop on carbon sequestration will hopefully avoid some of the pitfalls that limited commitment to and implementation of UNCOD's anti-desertification plan of action, although the absence of development-oriented organizations from the list of sponsors of this meeting is not an auspicious sign. In particular, it will be important to establish meaningful connections between scientific understanding of the problems and opportunities posed by efforts to deal with desertification issues and the practical realities of implementing field-level programs that can increase carbon sequestration so there will actually be something to monitor and verify.

Modeling Carbon Sequestration

Models are complicated scholarly products, difficult for the uninitiated to understand adequately let alone critique. But from the perspective of social science, the monitoring and verification models presented in the Post et al. paper raise some interesting issues. These issues are six in number:

1. How generalizable are the models? Most of them appear to be regional models based on data derived from North American grassland environments. The soil-forming processes in other drylands are unlikely to mirror these conditions, and certainly the socioeconomic environments in which resource management takes place are radically different. The vast majority of the experience of model creation and data collection referenced in the paper is North American in origin, and raises questions about how robust the models will be in estimating behavior of other physical systems.

2. All of the models cited deal with erosion. If this is the fundamental feature of C estimation models, there is a high degree of likelihood that C loss will be overestimated. Erosion is only one part of the total C balance in any area. Deposition is also an integral part of environmental change. While some areas will lose C through erosion, other areas will accumulate C through sedimentation. Because the models discussed assume reader familiarity and do not pres-

ent details of model design, it is possible that deposition is implicit in each model. But little evidence of the gainers from any erosion process in terms of the sinks to which the lost material is directed is found in the current version of the paper. Is it not possible that depositional zones are net beneficiaries in this erosion-based transfer of C? Is it possible that the carbon balance at a regional scale might be much more balanced than the losses of C noted at the field level might suggest?

3. If regional verification models are to be developed for areas outside the industrialized world, do data exist to put into the model? Or is a very expensive and time-consuming data collection effort needed to bring the verification model to operational status? While it is an act of hubris to assume that everything needed for success is available, it is equally problematic to assume that an ambitious, large-scale research agenda can be funded with traded carbon credits.

4. Is it possible to standardize SOC concentration measurements? The discussion of direct measurement points out the uncertainties involved in measuring SOC, but does not indicate what the most desirable measurement might be. Since the distribution of SOC can vary considerably from one soil to another, selection of a standard volume has potential for error. If it is possible to achieve an agreed measurement procedure, is it possible to construct a SOC map at a national scale where detailed data are lacking without engaging in a massive research effort?

5. Is it realistic to limit consideration of stored C to soil alone? In many dryland ecosystems, considerable amounts of C are locked up by net primary productivity (NPP) and are found in the vegetation, and relatively little organic material is present in the soil. Much of this NPP is found in annual rather than perennial vegetation, and heavy grazing pressure also removes much of the material that ordinarily would enter into the soil through decomposition. In such situations, calculations of C would underestimate production because it would be transported elsewhere either as fodder for consumption or on the hoof. In any case, viewing the local environment in which C is sequestered as a plant-soil system might be a more realistic way to envisage the carbon cycle.

6. Can models based on field data be built up to give an accurate picture of regional SOC? Interpolating between a small number of data points to construct a regional model is possible if spatial statistics are employed. Building up a model in this fashion is a difficult task, particularly in drylands where variability in resource potential is highly variable in time and space. Great care must be exercised in order to avoid overlooking significant local niches.

Questions Meriting Consideration

Dry environments are extremely variable places on both temporal and spatial scales. The soil-forming processes in these habitats are equally variable. How to incorporate this spatial variability into the models under discussion is a critical concern. Without this sense of spatial variation, the model, perhaps unjustly, conveys a static quality to the observer. Moreover, while the paper clearly shows that C status can be measured accurately on millennial time scales, it is less clear that measurements at shorter time scales are feasible. Accurate annual measures of SOC are probably not attainable or necessarily desirable, but five- to ten-year assessments would appear to be essential. If polluting nations trade for carbon rights, they are going to insist on knowing whether their investments are eliciting real gains in carbon sequestration!

That SOC parameters in North American soils are reasonably well understood is demonstrated in the Post et al. paper. But in non-industrialized countries with carbon rights to trade, it is not clear that soil processes will approximate those found in North America. One such parameter is the role played by fire in many tropical and subtropical grasslands and shrublands. The practice of burning as a management tool is widespread in Sahelian countries, and is encountered in many other places. When combined with naturally initiated fires, the impact on the local ecology can be considerable. It is not clear how fire affects C sequestration, since burning releases considerable C to the atmosphere but also returns C and other minerals to the soil. Fire is integral to the evolutionary development of the complex mosaic of vegetation found in many grassland ecosystems, and often is as important in determining the distribution of woody and grassy vegetation as is regional climate.

Linking Models to People

Ultimately the goal of resource assessment models is better management of the basic resources available to local communities. In the context of carbon sequestration, this means finding ways to increase the amount of carbon captured by the agro-pastoral systems employed by indigenous resource users. Traditional management systems, through long association with and survival in place, generally have a sound understanding of how to operate successfully. In the vast majority of cases, developing regional policies of land use that favor carbon accumulation must begin with what local people know, understand, and do at the field level. The Post et al. "strawman" plan to monitor SOC changes presumes that experts (presumably non-local) have the knowledge and technology to suggest appropriate C enhancing land use practices, and that teams of land managers, soil experts, economists, and industry representatives will impart this wisdom, consulting local technocrats and farmer groups primarily about the selection of study plots and the extrapolation of model results. This methodology is guaranteed to marginalize local expertise and to replicate the primary cause of development failures with which the African landscape is littered.

Comparing new and traditional management areas to show how much C is *not* lost in the new areas is an interesting technique. But it is as likely to give a false understanding of the local C status as it is to accurately measure improvements taking place on the "new" site. The fundamental assumption is false in many instances, because often the problems encountered by resource users in "traditional" areas are a direct result of the activities of "new" management. Resources devoted to improved areas under new management frequently extract resources critical to the success of traditional management from traditional areas. Forced to operate on poorer and more limited resources, it is small wonder that the agro-pastoral ecologies of traditional zones exhibit signs of stress (and low C status as well). In effect, the gains attained in new management areas occur at the expense of spatially more distant zones. Those areas remaining under traditional management often are "sacrificed" to the gains made elsewhere. Without addressing these issues of spatial inequality, without viewing the modeling activity in a holistic environmental and social context, models of carbon sequestration are apt to give a false picture of regional gains and losses in C status. Land use and land cover are closely linked phenomena, but the connections between land use activities and cover in one place and the changes occurring in other locales are often difficult to disentangle. But identification of these connections will be critical to determining whether and why the carbon status of local soils is changing and in what direction the regional change is taking place.

Most of the niches that are most likely to accumulate large amounts of C are wet sites that are already being heavily used by local management systems. Displacing these systems by "new" management schemes will have very serious consequences. Upgrading more traditional systems by introducing new techniques and technology is not a simple task; rather, it requires a long-term commitment to gradual change that blends the new and the old and grows the amalgam steadily into altered (and hopefully improved) states. Technology introduced from the outside without careful understanding of the local social and ecological setting has a tragic record of project failure, undesirable social impacts, and environmental damage. Much of this negative change could have been avoided if the lure of rapid productivity gains had been moderated by more realistic assessments of what was attainable. For example, unregulated introduction of deep bore wells into water deficit pasture areas in the Sahelian zone is probably the single most important factor leading to rangeland degradation. With access to water open to all, traditional constraints on pasture use were abandoned and degradation in range quality followed inexorably. Thus any search for new ways to promote C storage either by introducing new techniques into existing systems or by the areal extension of new systems into lower intensity use areas must proceed from a position of extreme sensitivity to the current indigenous social system and its management techniques.

Moreover, it is by no means certain that the traditional regimes do a bad job of land use management. Indeed, in many part of the Sudanese Sahel, to cite but one example, the traditional patterns of common property rangeland grazing management, together with an agricultural system of frequently shifting fields, long fallows, and gum arabic extraction from *Acacia senegal* that spontaneously regenerated on fallow plots undoubtedly resulted in higher levels of SOC than the current pattern of more intensive, machine-dominated cereal cultivation. While it is not possible to turn the clock back to a nostalgic, pristine prior social and environmental state, it is feasible to build on local principles of resource management.

An extremely important feature of any SOC model-building effort should be the identification of those local areas where carbon sequestration is most likely and of the local actions that support carbon accumulation either in soil or vegetation.

These places and practices need to be identified, carefully examined, and then expanded spatially. Wherever possible, local self-help initiatives and institutions, far more numerous than are often credited by outside observers, should constitute the basic building block for the investment of carbon transfer dollars. Building on the indigenous experience and ethno-scientific wisdom of local agro-pastoral ecologies is the path least trodden in development but the one most susceptible of long-term success. The need is not for large-scale, new research programs with elaborate experiment stations, but rather for a new partnership between science and local knowledge systems that incorporates local ways of knowing and doing as the springboard for positive environmental change.

Conclusion

Models can be constructed that aid in verifying and monitoring the C status of environments at local, regional, national, and global scales. Constructing models that are robust and that realistically establish linkages between scales is an extremely challenging task. Incorporating traditional indigenous management systems and knowledge into such models, rather than relying solely on big science for solutions, is a gigantic undertaking, fraught with difficulties but susceptible of unique and enduring results. Socializing the pixel by including local social systems in fundamental ways and by basing monitoring systems on indigenous expertise is critical to the success of carbon sequestration efforts.

HERMAN MAYEUX, *United States Department of Agriculture*

The authors describe the paper as a discussion of current methods for measuring carbon changes in soil and an analysis of sources of error arising from application of those methods in monitoring and verification schemes. They also attempt to "lay out the basis for a verification and monitoring plan." They succeed in the first two goals, describing the potential roles of direct sampling (of stocks and fluxes), remote sensing, and modeling, and the inherent constraints associated with each. The authors draw needed attention to problems of scale, spatial and temporal heterogeneity, and limits to detection of changes. The discussion of potential values of process and accounting models is useful and well done, and the assessment of the usefulness of stratified accounting models that employ GISs, geographic databases, and other inputs raises several interesting considerations, many of which were new to me.

Options for improving the utility of stratified accounting models by "combining spatially detailed information" to avoid errors in spatial averaging were discussed. An omitted option with possible potential may be the use of direct measurements, especially those from comparisons of SOC levels from experimental manipulations and in situ comparisons that were maintained long enough (on the order of 5 years) to allow SOC responses to be detected. Decades of such research have resulted in hundreds of publications which provide quantitative SOC responses to soil management alternatives and conservation practices, often presented as changes relative to adjacent areas maintained for long periods under traditional or conventional management. In the aggregate, results of these studies may be more useful than some of the attributes and ancillary data identified in the paper. Certainly, the availability of direct measurements of SOC changes in response to management is increasing rapidly because of the current rise in interest in the effects of tillage and cropping systems and conservation practices on soil quality.

The third objective of the paper is clearly a difficult one, whether phrased as defining a basis for verification or presenting a "strawman" plan. The authors begin by proposing that the first step that might be taken "in developing methods to verify and monitor SCO changes at a given scale" is to identify the areas within a state or country that are most responsive to carbon management, using management history, consultations, remotely sensed data, and other databases. The reason for this was not clear. Perhaps this should be done if the goal is to maximize C storage. To develop methods, we might begin in areas where we have sufficient direct measurements of SOC responses to human activities to provide calibration and validation.

Figure 2-5 represents a conceptual representation of landscape components and tools needed to monitor C dynamics, and it may represent such an area. The suggestion here is that changes in SOC would be monitored as changes in C pools and fluxes at the landscape level, an area including different land uses, both woodland and cropland in this example. Approaches attempted or proposed to date operate at the level of ecosystem types or individual land uses, which makes sense if, in addition to monitoring changes in C stocks as defined in the Kyoto Protocol, we view the monitoring process as containing a verification component for rewarding managers (like farmers) of landscape components (like fields) for storing C. The landscape approach may be necessary to attain the ultimate goal of developing "macrotools" for making large scale estimates, but it seems that a focus on the landscape scale requires some discussion of methods to get to that point from ecosystem types or specific land uses, such as cobbling together of smaller scale accounting or process models.

In Figure 2-5, the authors propose a C balance equation that accounts for changes in SOC at time t as the sum of SOC at time zero. Components include photosynthetically assimilated C, C buried in the erosional process, and C lost to harvest of biomass, respiration, and erosional displacement over time t. Components in the figure suggest that the equation is balanced by a combination of direct analysis of soil samples, by eddy flux measurements, and by satellite observations. Geographical Information Systems (GISs), databases, and simulation models also appear. The figure incorporates most known options for making the necessary measurements and observations, and the discussion that follows provides detailed information and assessments of the current status of those options for quantifying the terms in the equation. It may come as close to a strawman plan as we can get at this point, but it is not clear how this approach addresses problems of spatial heterogeneity and others described in the paper.

The authors briefly consider the use of simplified models like a "soil carbon equation" and indicate that these might have value under specified soil/climate conditions. This seems completely true. But, at least in the case of agricultural land uses, it is possible to develop a matrix of soil carbon equations that includes essentially the entire spectrum of specific soil/climate conditions for the United States, where we have a complete and rather detailed soil survey. Because soil taxonomy reflects climate, it may be that only soil taxonomy need be considered in such a matrix. Consider a table with rows representing associations of similar soil series (associations are already defined) or even further aggregations of soil series, perhaps based on texture, and columns representing human actions or causes of SOC change, such as enrollment in the Conservation Reserve Program or the shift from conventional to one of the conservation tillage systems. Each cell in the table could contain a simple equation representing a given soil association's response to an action. If climate and other geographical variables matter, such a table could be constructed for areas such as the ecoregions used for model comparisons and other applications, or the MLRAs or LRAs (Land Resource Areas) already in use by USDA. The simple equation might yield changes in SOC expressed as mass per unit area per year, calibrated by field measurements of SOC change from the literature and additional in situ analyses, where required.

The authors point out, correctly, that changes in SOC content can be determined from repeated soil C analyses and from flux measurements, and that the accuracy and cost of flux measurements made by eddy flux technology will improve as experience is gained over time. It is also likely that new technology for making inexpensive but accurate flux measurements will soon be available. For example, the conductometric method, based on changes in the electroconductivity of water with CO_2 concentration, has recently become quantitative to less than 0.1 ppm, at least in the differential mode, with response time reportedly down to 30 s. A pair of such sensors costs only $300 if ordered from a scientific supply company, but most shops can make it for much less.

SHASHI B. VERMA, *University of Nebraska*

This paper is well written and contains useful information on soil properties and processes, and techniques for measuring and modeling soil C changes. Brief mention has been made of the eddy covariance technique for measuring the net ecosystem CO_2 exchange. In the following I include comments on a few specific points discussed in the paper. I strongly recommend that the eddy covariance method be given move emphasis in the paper. I believe the method can be used for continuous measurements of C uptake by the whole ecosystem, and these measurements, in conjunction with the soil C measurements, will provide more comprehensive information on C sequestration.

Specific Comments

- I believe the eddy covariance method is based on fundamental principles of atmospheric turbulence and is a direct method (as are the other methods referred to in the paper). This method causes minimal disturbance to the micro-environment being studied, and provides continuous measurements of carbon flux.

- It would be useful to provide a few selected references which discuss the basic principles involved (e.g., Businger, 1986; Baldocchi et al., 1988; Verma, 1990).
- Table 2-3: The term "experimental" seems to have a negative connotation. The eddy covariance method has been used for about 20 years for CO_2 flux measurements (and for several years for continuous, year-round measurements—e.g., Wofsy et al., 1993; Goulden et al., 1996a,b, 1998). This method is presently being employed at about 60 stations worldwide for continuous, long-term CO_2 flux measurements.
- Table 2-3: I think we need to be careful about expanding the application of this method to automated weather stations. Many of these weather stations may not provide satisfactory "upwind fetch" needed for accurate eddy covariance flux measurements (e.g., Schuepp et al., 1990; Horst and Weil, 1992, 1994).
- Table 2-3: In view of the comment given above, I would recommend an approach which utilizes a coupling of the eddy covariance flux measurements (made at suitable sites) with appropriate terrestrial biosphere models (e.g., SiB2: Sellers et al., 1996). After such models have been thoroughly tested in relevant ecosystems using the eddy covariance flux measurements, they can be applied to various locations to accurately calculate the net ecosystem CO_2 exchange.

References

Baldocchi, D.D., B.B. Hicks, and T.P. Meyers. 1988. Measuring biosphere-atmosphere exchanges of biologically related gases with micrometeorological methods. *Ecology* 69(5):1331-1340.

Businger, J.A. 1986. Evaluation of the accuracy with which dry deposition can be measured with current micrometeorological techinques. *J. Clim. And Appl. Meteorol.* 25:1100-1124.

Goulden, M.L., J.W. Munger, S.-M. Fan, B.C. Daube, and S.C. Wofsy. 1996a. Exchange of carbon dioxide by a deciduous forest: response to interannual climate variability. *Science* 271:1576-1578.

Goulden, M.L., J.W. Munger, S.-M. Fan, B.C. Daube, and S.C. Wofsy. 1996b. Measurements of carbon sequestration by long-term eddy covariance: methods and a critical evaluation of accuracy. *Global Change Biology* 2:169-182.

Goulden, M.L., S.C. Wofsy, J.W. Harden, S.E. Trumbore, P.M. Crill, S.T. Gower, T. Fries, B.C. Daube, S.-M. Fan, D.J. Sutton, A. Bazzaz, and J.W. Munger. 1998. Sensitivity of boreal forest carbon balance to soil thaw. *Science* 279:214-217.

Horst, T.W. and J.C. Weil. 1992. Footprint estimation for scalar flux measurements in the atmospheric surface layer. *Boundary-Layer Meteorol.* 59:279-296.

Horst, T.W. and J.C. Weil. 1994. How far is far enough? The fetch requirements for micrometeorlogical measurement of surface fluxes. *J. Atmospheric and Oceanic Technology* 11:1018-1025.

Schuepp, P.H., M.Y. Leclerc, J.I. MacPherson, and R.L. Desjardins. 1990. Footprint prediction of scalar fluxes from analytical solutions of the diffusion equation. *Boundary-Layer Meteorol.* 50:355-373.

Sellers, P.J., D.R. Randall, G.J. Collatz, J.A. Berry, C.B. Field, D.A. Dazlich and G.D. Colello. 1996. A revised land-surface parameterization (SiB2) for GCMs. Part 1: Model formulation. *J. of Climate* (in press).

Verma, S.B. 1990. Micrometeorological methods for measuring surface fluxes of mass and energy. In N. S. Goel and J. M. Norman (eds.), *Instrumentation for Studying Vegetation Canopies for Remote Sensing in Optical and Thermal Infrared Regions.* Special Issue of Remote Sensing Reviews 5(1):99-115.

Wofsy, S.C., M.L. Goulden, J.W. Munger, S-M. Fan, P.S. Bakwin, B.C. Daube, S.L. Bassow, and F.A. Bazzaz. 1993. Net exchange of carbon dioxide in a mid-latitude forest. *Science* 260:1314-1317.

3

Desertification Control to Sequester C and Mitigate the Greenhouse Effect

R. Lal, *Ohio State University*

H.M. Hassan, *The World Bank*

J. Dumanski, *The World Bank*

The word *desert* or *tesert* is an ancient Egyptian hieroglyph meaning a place that was abandoned, forsaken or left behind (El-Baz, 1988), from which is derived the Latin verb *"deserere"* which means "to abandon." Other derivatives from these roots are such words as *desertum* (a waste place or wilderness), and *desertus* (abandoned or relinquished). Therefore, the word *desert* implies that it was a better place but is now wasted and has been abandoned. The term desertification was first used by a French forester in 1949 (Auberville, 1949). He used the term to describe a general process of degradation starting with deforestation (not necessarily in drylands) and ending in lands turned into desert. His work was based in humid tropical West Africa. From this first beginning, the term is now widely used to mean the spread of desert-like conditions (Mainguet, 1991). It implies more than just shifting sand dunes, because it may occur thousands of kilometers away from margins of true deserts. Although desertification mostly occurs in drylands, it is not limited to these regions. Several other terms have been used to describe the process, including "aridization," "aridification," "xerotization," "desert encroachment," and "desert expansion." However, the term "desertification" is all encompassing, conveys the meaning of the complexity of all processes involved, and refers to drought, erosion and denudation of vegetation cover leading to spread of desert-like conditions.

The literature, both scientific and popular, contains over one hundred definitions of the term (Glantz and Orlovsky, 1986: Odingo, 1990). UNEP (1977) defined desertification as "the diminution or destruction of the biological potential of land which can lead ultimately to desert-like conditions." The most comprehensive explanation of the process is that by Verstraete (1986) who stated that "desertification is neither drought, nor soil erosion, nor the destruction of the vegetation cover, nor the cutting of trees, nor even the degradation of living conditions alone. It is all that and much more." For the objective of assessment of land area affected, UNEP (1990) and UNCED (1992) defined desertification as "the land degradation in arid, semi-arid and dry sub-humid areas resulting from various factors including climatic variations and human activities." The term land includes whole ecosystems comprising soil and water resources, land surface, vegetation, animals and crops. The term degradation implies reduction of resource potential by one or a combination of processes, e.g., water and wind erosion and sedimentation, long-term reduction in the amount and diversity of natural vegetation and animals, and salinization. These definitions, despite being comprehensive, do not cover the problem of desertification in higher latitudes (e.g., Iceland) and need to be revised and made inclusive.

Desertification may be regarded as a "process" or the ultimate "stage" of land degradation. The latter refers to the "point" when land becomes unproductive and has severe economic limitations (Mainguet, 1991). In general, the term desertification refers to the "process" of land degradation in arid, semi-arid and dry sub-humid areas resulting mainly from adverse human impact (Mendoza, 1990; Middleton and Thomas, 1992). However, in a broader context, desertification implies adverse alterations in all aspects of the natural or biophysical environment by human actions, to the detriment of vegetation, soils, landforms, water and ecosystems (Conacher, 1998), and human well-being. Although the problem of desertification may occur in all climates and ecoregions, this chapter focuses on the problem of desertification only in drylands of the world.

The objectives of this chapter are to (a) collate and synthesize the available literature regarding the impact of desertification on soil organic carbon (SOC) pool and fluxes, (b) infer the effect of desertification control and land restoration on enhancement of soil quality, and (c) assess the potential of desertification control to sequester C in soil and mitigate the greenhouse effect.

WORLD'S DRYLANDS

Drylands comprise a wide range of ecoregions from hyper-arid to sub-humid. These lands are characterized by low and erratic rainfall/precipitation. Typically, **hyper-arid** areas receive <200 mm of precipitation annually, **arid** areas receive less than 200 mm of winter rainfall or less than 400 mm of summer rainfall, and **semi-arid** areas 200 to 500 mm of winter rainfall or 400 to 600 mm of summer rainfall, and dry sub-humid regions from 500 to 700 mm of winter rainfall or 600 to 800 mm of summer rainfall (FAO, 1993). In addition to precipitation, drylands can also be defined on the basis of the length of the growing season (days) as determined by the favorable water balance and temperature regime. Arid lands typically have <75 growing days, semi-arid lands have 75 to 120 days and dry sub-humid regions from 120 to 150 days. The principal feature of these lands is the negative water balance, as expressed by the difference between annual rainfall/precipitation and potential evapotranspiration.

The water balance in drylands is influenced by precipitation, evapotranspiration, topography/landscape and soil profile characteristics. However, most indices used to classify drylands are based on climatic factors and do not include soil characteristics and topographical features. Two commonly used indices are the dryness ratio of Budyko (1958) (Table 3-1) and the Aridity Index. The most widely used Aridity Index is defined as the ratio of the precipitation (P) to the potential evapotranspiration (PET). The ratio is less than 0.05 for **hyper-arid** regions, 0.05-0.20

Table 3-1. Dryness ratio in relation to vegetation zones of sub-Saharan Africa (Hare, 1985; Lettau, 1969; Budyko, 1958).

Dryness ratio (D)	Vegetation zone
>10	True desert
7-10	Sahelo-Saharan
2-7	Sahelian and Sahelo-Sudanian
1-2	Sudano-Guinean and Sudanian
<1	Guinean

D = R/LP, where R is annual net radiant energy balance at earth's surface, L is the thermal energy required to evaporate a year's rainfall, P is average annual rainfall.

for **arid** regions, 0.20-0.50 for **semi-arid** regions and 0.50-0.65 for **dry sub-humid** regions. Using these criteria, the total land area of the earth in arid regions is 61.50 $\times 10^6$ km^2 or 47.1% of the earth's land area distributed in all continents (Table 3-2). If the hyper-arid regions are excluded, as per UNEP (1992) definition, then total land area prone to desertification is 51.72 $\times 10^6$ km^2. Lack of or low level of precipitation in these regions is caused by anticyclonic subsidence, leeward locations, continental locations, and oceanic cold currents (Noin and Clarke, 1997). The global land area covered by different zones is 7.5% by **hyper-arid**, 12.0% by **arid**, 17.7% by **semi-arid** regions and 9.9% by **dry sub-humid** regions (Figure 3-1). At the generalized scale presented in Figure 3-1, there is a considerable uncertainty ($\pm$ 10% or more). In particular, the map does not include steep lands of Central America, Haiti, Caribbean Islands and regions in higher latitudes.

Table 3-2. Arid regions of the world as per the aridity index (modified from Middleton and Thomas, 1992: Noin and Clarke, 1997).

Aridity Index	Hyper-arid <0.05	Arid 0.05-0.20	Semi-arid 0.20-0.50	Dry sub-humid 0.50-0.65	Total	% of Earth's area
Region	10^6 km^2					
Africa	6.7	5.0	5.1	2.7	19.6	15.0
Asia	2.8	6.3	6.9	3.5	19.5	14.9
Australasia	0	3.0	3.1	0.5	6.6	5.1
Europe	0	0.1	1.1	1.8	3.0	2.3
N. America	0.03	0.8	4.2	2.3	7.4	5.6
S. America	0.3	0.5	2.7	2.1	5.4	4.2
Total	9.8	15.7	23.1	13.0	61.5	
% of Earth's land area	7.5	12.0	17.7	9.9	47.1	

Aridity Index = Precipitation/Evapotranspiration
1 km^2 = 100ha

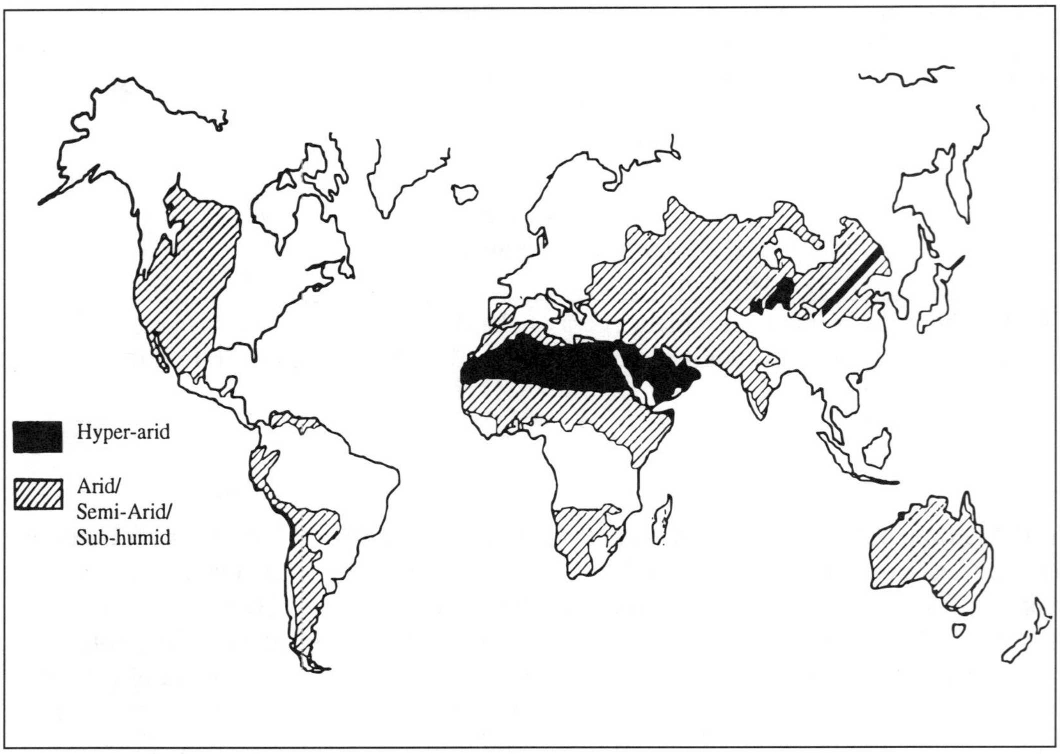

Figure 3-1. Dryland areas of the world (Glenn et al., 1993).

Land degradation, a serious problem in drylands (Whitehead et al., 1985; Barrow, 1991; Beaumont, 1989; Dregne, 1995, 1998), results from a combination of soils and environmental factors compounded by social and political conditions (Darkoh, 1998). Soils of the drylands also vary widely (Table 3-3). Principal soils comprise young soils of coarse-texture, low organic matter content, low water and nutrient retention capacities, and low inherent soil fertility. Perkins and Thomas (1993) reported that SOC content of soils of the Kalahari Desert of Botswana ranged from 0.2 to 0.6%. Balpande et al. (1996) reported that SOC content of Vertisols in central India ranges from 0.1 to 0.4%. Young and poorly developed Entisols and Aridisols comprise 75% of all soils. Relatively fertile soils (Mollisols, Alfisols and Vertisols) constitute about 22% of the total land area. High potential soils, with a capacity to produce large biomass under optimal conditions, are rather rare in these regions.

Drought stress is the principal constraint to biomass production. Therefore, arable land use is restricted only to semi-arid and dry sub-humid regions or where irrigation is available. Nomadic pastoralism and ranching are the key land uses in hyper-arid and arid regions (Heathcote, 1983).

Table 3-3. Principal soils of the arid regions (10^6 km^2) (recalculated from Dregne, 1976; Noin and Clarke, 1997).

Soil	Arica	Asia	Australia	N.America	S.America	Total
Alfisols	2.351(12)	—	.464(7)	.294(4)	.706(13)	3.815(6.2)
Aridisols	5.485(28)	7.991(41)	2.917(44)	3.312(45)	1.520(28)	21.225(34.5)
Entisols	11.362(58)	6.627(34)	2.453(37)	.589(8)	2.281(42)	23.312(37.9)
Mollisols	.196(1)	3.898(20)	—	3.018(41)	.923(17)	8.035(13.1)
Vertisols	.169(1)	.975(5)	.796(12)	.147(2)	—	2.114(3.4)
Total	19.59	19.491	6.630	7.360	5.430	

Figure in parentheses is % of the land area in that soil class.
This table does not include soil distribution in 3.00 10^6 km^2 of arid lands in Europe.

EXTENT AND RATE OF DESERTIFICATION

Since the 1970s increasing demographic pressure in these harsh climates with fragile ecosystems have focused global attention on these ecoregions. However, land degradation and desertification are not new; they have plagued mankind since the dawn of civilization. Salinization and siltation affected the Lower Mesopotamian irrigation scheme from about 2400 BC (Jacobsen and Adams, 1958). Wind and water erosion and salinization were the problems of the Indus Valley civilization since 3000 to 2500 BC (Durrans and Knox, 1982). Salinization and erosion have been the common causes of the downfall of numerous past civilizations of the arid regions (Thomas and Middleton, 1994), including those in central Asia (present day Uzbekistan) (Maingeut, 1991), China's northwestern Tarim Basin (Zhao, 1986), and the Negev Desert (Issar and Tsoar, 1987). The problem of desertification, however, has been accentuated by increasing demographic pressure (Zaimeche and Sutton, 1990; Lewis and Nyamulinda, 1996), land misuse and soil mismanagement, and attendant severe stress on the fragile environment leading to adverse effects on soil, water, range, forest, livestock and wildlife (Darkoh, 1990). In Kuwait, for example, agricultural development based on irrigation has caused severe problems of salinity, soil erosion and sand encroachment (Omar et al., 1998). Similar problems have been experienced in Sudan, Egypt and elsewhere in the Mediterranean region. In China, more than 150 million hectares (Mha) are presently affected by accelerated wind and water erosion (Higgit and Rowan, 1996). The Ethiopian Highlands are losing soil at an alarming rate due to high demographic pressure (Wakeel and Astartke, 1996). In Niger, Chappell et al. (1998) estimated that the net soil erosion in semi-arid regions with rainfed agriculture may be 26 to 46 Mg/ha/yr, with some fields losing as much as 48.5 Mg/ha/yr.

Exact estimates of the land area affected by desertification are difficult to obtain because of the lack of quantitative criteria in defining desertification and of suitable methods in assessing it. UNEP has obtained three separate estimates ranging from 3.5 to 4.0 billion hectares of land area desertified to some degree (Table 3-4). The problem is also confounded by discrepancies in other estimates (Table 3-5), because the criteria used are different. Mabbutt's (1984) estimates are low because of the inclusion of severely affected areas only. The GLASOD data are also based on a subjective methodology (Oldeman and Van Lynden, 1998) and are variable

Table 3-4. UNEP estimates of land area susceptible to desertification.

Parameter	UNEP (1977)	UNEP (1984a, GAP-I)	UNEP (1992, GAP-II)
Total area (Mha)	5550	4409	5160
% of area susceptible to desertification	71.5	78.9	69.6
Total area susceptible to desertification (Mha)	3970	3475	3592

Table 3-5. Estimates of area affected by desertification.

Region	FAO/UNESCO/WMO (1977)	Dregne (1983)	Mabbutt* (1984)
	$10^6 km^2$		
Africa	13.7	10.7	7.4
Asia	10.4	12.0	7.5
Australia	5.7	3.1	1.1
Europe	0.2	0.2	0.3
North & Central America	4.3	3.2	2.1
S. America	3.3	3.3	1.6
Total	37.6	32.5	20.0

* Severely affected

(Table 3-6), but most reliable among all estimates. According to the GLASOD methodology, the land areas subjected to soil degradation in arid regions range from 1016 to 1137 Mha, with about 139 Mha subjected to severe and extreme degradation. UNEP (1991) made a distinction between land degradation and vegetation degradation. Vegetation in rangeland can undergo degradation with or without soil degradation. The data in Table 3-6 show that area affected by land degradation is 1016 Mha, and an additional area of vegetation degradation is 2576 Mha.

Similar to the area affected, estimates of the current rate of degradation also vary widely. The rates of desertification estimated by UNEP have also varied from 21 Mha/yr (UNEP, 1984b) to 27 Mha/yr (UNEP, 1987). Mainguet (1991) estimated the annual rate of desertification at about 5.8 Mha with 55% of it occurring on rangelands and 43% on rainfed cropland (Table 3-7). Relatively, the rate of degradation is, in decreasing order: rainfed cropland (0.439%/yr) > irrigated land (0.095%/yr) > rangeland (0.086%/yr) (Table 3-7).

Table 3-6. GLASOD estimates of desertification (e.g. land degradation in dry areas excluding hyper-arid areas).

UNEP (1991)		Oldeman and Van Lynden (1998)	
Land type	Area 10^6km^2	Type of soil degradation	Area 10^6km^2
Degraded irrigated lands	.43	Water erosion	4.78
Degraded rainfed cropland	2.16	Wind erosion	5.13
Degraded rangelands (soil and vegetation)	7.57	Chemical degradation	1.11
		Physical degradation	.35
Sub-total	10.16		
		Total	11.37
Degraded rangeland (vegetation alone)	25.76	---------------------------	
		Light	4.89
Total	35.92		
		Moderate	5.09
Total land area (excluding hyper-arid regions)	51.72	Severe and extreme	1.39
% degraded	69.5	Total	11.37

The estimate by Oldeman and Van Lynden does not include the vegetation degradation on rangeland.

Table 3-7. Estimate of annual rate of land degradation (calculated from Mainguet, 1991; UNEP, 1991).

Land use	Total land area (Mha)	Rate of land degradation (Mha/yr) Mha/yr	% of total/yr
Irrigated land	131	0.125	0.095
Rangeland	3700	3.200	0.086
Rainfed cropland	570	2.500	0.439
Total	4401*	5.825	0.132

*This figure of total land area exceeds that shown in Table 3-4 (3970 Mha) and represents variations in different estimates of the total susceptible area.

GREENHOUSE EFFECT AND DESERTIFICATION

Because of the increase in atmospheric concentration of CO_2 and other radiatively active gases (CH_4, N_2O, CFCs), it is argued that the global mean temperature may increase by 0.8 to 4.1C between 1980 and 2030 (IPCC, 1996). However, some areas may warm more than others. Some may become drier and others wetter. There is likely to be a northward shift in the mid-latitude rainbelt (Gleick, 1985). The warming effect is expected to be lower over the ocean than over land and increase from the Equator to the Poles. Temperature increase in the drylands is, therefore, likely to be smaller than that for areas at medium and high latitudes (Grainger, 1990).

Several uncertainties impede an objective description of the impact of greenhouse warming on desertification. On the one hand, causes of desertification are numerous, including land misuse and mismanagement attributable to high demographic pressure, natural climatic cycles, and combinations of both. On the other hand, the issue is further confounded by the uncertainties in projected global warming, especially in the dry areas of the world.

Despite complexities and uncertainties, several attempts have been made to predict the impact of the greenhouse effect on desertification (IPCC, 1996; Houghton et al., 1990; Grainger, 1990). Ojima et al. (1993) estimated that C pool in drylands may increase due to better moisture regimes and the CO_2 fertilization effect on biomass production. They estimated an increase in the net sink capacity from 5.6 to 27.4 Pg C by 2040 due to increase of 280 Mha in area of the warm grassland sub-biome. This would be caused by biome shift due to the projected climate change and sustainable management of grasslands. On the contrary, regressive land management may lead to C emission at the rate 0.7 Pg C/yr with a potential loss of 11 to 37 Pg C over the 50-year period from 1990 to 2040 (Ojima et al., 1993). Most available statistics such as these, however, are derived from models based on numerous assumptions that may be hard to validate, and none of the models considers the potential of C sequestration through formation of secondary carbonates.

DESERTIFICATION IMPACT ON SOIL QUALITY AND HISTORIC LOSS OF SOIL C

Desertification leads to decline in soil quality. Some important indicators of decline in soil quality are reduction in soil structure, the depletion of soil organic carbon (SOC) pool, and loss of biodiversity (SSSA, 1995; Smyth and Dumanski, 1993). The principal processes of land degradation, which also exacerbate desertification, include erosion by water and wind, soil compaction and decline in soil structure, salinization, fertility depletion and nutrient imbalance, and decline in soil biodiversity. All these processes lead to decline in soil quality, reduction in above- and below-ground biomass production, disruption in the C cycle, and reduction in the SOC pool.

Several long-term experiments in the dry areas have shown that continuous cropping without adequate levels of fertilizers and other recommended input can lead to decline in soil structure and SOC content. The SOC content of soils of the arid zone is usually low (Kolarkar et al., 1994: Dhir and Mann, 1977), and it declines with cultivation. In northwestern Nigeria, Raji et al. (1996) observed that soils of the stabilized sand dunes are extremely low in their levels of SOC, often less than 2 g/Kg. Swift et al. (1994) showed that continuous cultivation for 14 years without recommended inputs of fertilizers and manures halved SOC content (from 2% to 1%) of soils in East Africa. The SOC content was maintained at about the antecedent level with application of fertilizers at the recommended rate, use of farmyard manure and return of crop residue to the soil surface. The data from several long-term experiments conducted in Francophone sub-Saharan Africa have been summarized by Pieri (1991). These data showed that continuous cropping without application of fertilizers and/or manure leads to a rapid decline in SOC content. Pieri developed a nomogram relating critical level of SOC content required to maintain physical stability for soils of different textural composition. In the Mallee area of southeastern Australia, Leys and McTainsh (1994) observed that wind-blown sediments were enriched in SOC, N and P contents. The enrichment ratio of sediments was 16 for SOC and 11 for total N. The high enrichment ratio of sediment indicates severe decline in soil quality on site where erosion occurred.

Assessing the impact of desertification on historic loss of soil C is a highly challenging task. The soil C pool comprises two components, SOC and soil inorganic carbon (SIC) pools. The SOC pool, primarily confined to the upper 30 to 50 cm of soil profile, is low in soils of dry areas and declines rapidly with traditional methods of cultivation. Loss in the SOC pool is accentuated by land degradation due to erosion and other processes. With some exceptions, there is a scarcity of experimental data relating past land use to the SOC pool. It is difficult to obtain reliable estimates of historic C loss from extrapolation of a few point measurements of SOC pool to billions of hectares of dryland in diverse ecoregions. The SIC pool (total lithogenic and pedogenic) is not widely studied. Its dynamics, in relation to land use and management, is even less well understood than that of the SOC pool. Therefore, the available information on the soil C pool and fluxes in dry areas, and the historic loss of C due to past management is sketchy and tentative at best.

Ojima et al. (1993) estimated that grasslands and drylands of the world have lost 13.1 to 23.6 Pg C due to anthropogenic perturbations (or desertifications). The data in Table 3-6 show that estimates of desertification include 1137 Mha of land degradation and 2576 Mha of vegetation degradation. Assuming that land degradation has led to SOC loss at the rate of 8 to 12 Mg C/ha, the historic C loss is 9 to 14 Pg C. Similarly, if the vegetation degradation has led to historic C loss of 4 to 6 Mg C/ha, the historic C loss is 10 to 15 Pg C. Therefore, the total historic C loss due to desertification may be 19 to 29 Pg C. Assuming that two-thirds of the C lost can be resequestered (IPCC, 1995) through soil and vegetation restoration, the potential of C sequestration through desertification control is 12 to 19 Pg C. This potential may be realized over a 40- to 50-year period. These estimates provide us with a reference point regarding the historic C loss and potential of C sequestration through restoration of soil and biotic ecosystems in desertified lands.

SOIL EROSION AND C DYNAMICS IN DESERTIFIED AREAS

Accelerated soil erosion influences C pool and fluxes because of the breakdown of aggregates, exposure of C to climatic elements, transport of sediments rich in SOC downslope, and sequestration of C with sediments in depositional sites and aquatic ecosystems. In general, C content of water and wind-borne sediments is more than that of the contributing soil. In southwest Niger, Sterk et al. (1996) reported that the wind-blown material trapped at 2 m contained 32 times more C than the topsoil. The C content in the dust sample was 5.36% compared with 0.15% in the topsoil. The data in Table 3-8 show estimates of the impact of wind and water erosion on C dynamics. Assuming that 20% of the C displaced is oxidized and emitted into the atmosphere (Lal, 1995; Lal et al., 1998), erosion leads to emission of 0.262 Pg C/yr.

Table 3-8. Soil erosion and SOC dynamics.

Degree	Total area affected by wind and water erosion (Mha)	[a]Presumed rate of soil erosion		[b]Total amount of sediments displaced (10^9 Mg/yr)	[c]Total C displaced (10^9 Mg/yr)	[d]C emission to the atmosphere (10^9 Mg C/yr)[e]
		Multiple of T value	Rate (Mg/ha/yr)			
Slight	372.3	1.25	14	52.1	0.52	0.08-0.10
Moderate	423.9	1.5	17	72.1	0.72	0.11-0.14
Strong	97.0	2.0	22	21.3	0.10	0.015-0.02
Extreme	6.6	3.0	34	2.2	0.01	0.0015-0.002
Exposure of Calciferous Horizon	—	—	—	—		0.02-0.03
Total						0.227-0.292

Assumptions
[a]T value is 11.2 Mg/ha/yr (T refers to tolerable soil loss, estimated at 5 ton/acre which is 12.5 Mg/ha)
[b]A delivery ratio of 10%
[c]SOC content of 1% for slight and moderate erosion and 1.5% in sediments of strong and extreme erosion
[d]C emission at 15-20% of SOC displaced
[e]1 Pg = petagram = 10^{15} g = 10^9 Mg = 1 billion ton
[f]Strong and extreme erosion may lead to erosion of sub-soil with low SOC content. In such cases, the loss of C by erosion would be less.

In addition to SOC, erosion also leads to exposure of the sub-soil rich in calciferous materials (Plates 2 and 3). These areas, severely affected by strong and extreme wind erosion, are estimated at about 20 Mha. Estimates of the land area with calciferous horizon exposed to the soil surface are not available. Therefore, it is assumed that 50% of these areas or 10 Mha have a calciferous horizon exposed at the soil surface. The exposed calciferous horizon is subject to the impact of anthropogenic perturbations and climatic factors (e.g. plowing, application of fertilizers, root exudates, acid rain etc.). These factors may lead to dissolution of carbonates and emission of CO_2. If this exposed layer containing high amounts of carbonates and bicarbonates leads to emissions of C at the rate of 20 to 25 g/m^2/yr, the annual rate of emissions of C from SIC is 2.5×10^6 Mg C/yr. Therefore, total C emission due to soil erosion is 0.227-0.292 Pg C/yr (Table 3-8).

STRATEGIES FOR DESERTIFICATION CONTROL AND LAND RESTORATION

Drylands by their very nature are resource-limited—limited in water and plant nutrients. That is why less than 10% of drylands are cropped, and a large part of rangeland is extremely arid and severely constrained by drought. Thus, productivity of drylands under traditional management systems based on low external input is extremely low. Consequently, key strategies for sustainable development, based on desertification control and land restoration are (1) arrest land degradation, and (2) enhance food security through alleviation of drought stress. Achieving food security and desertification control go hand-in-hand. Four important strategies for achieving these are (Figure 3-2): (1) enhancing water use efficiency, (2) controlling erosion and restoring degraded lands, (3) managing and enhancing soil fertility, and (4) increasing production of biofuels and fodder.

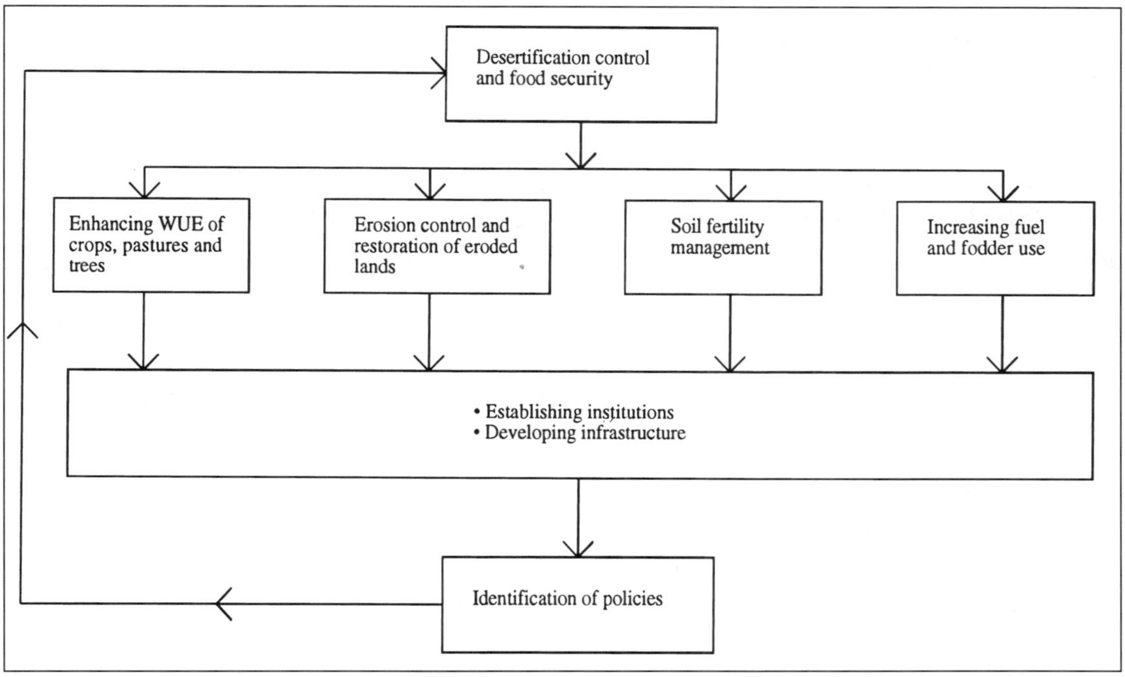

Figure 3-2. Strategies for desertification control and achieving food security (WUE=water use efficiency).

Appropriate Plant Species

Suitability of plant species for adaptation to drylands is governed by many adaptive mechanisms. Important among these are nature of the root system, growing season duration, canopy structure, leaf and stomatal characteristics. Most dryland plants close their stoma during daytime to reduce transpiration and conserve water. Such plants have, therefore, developed special photosynthetic mechanisms which are "add-ons" to the normal Calvin cycle or the C_3 mechanism (the photosynthetic pathway involves an intermediate 3-carbon molecule). These mechanisms include the C_4 (intermediate 4-carbon molecule) photosynthetic C fixation and the crassulacean acid metabolism (CAM) pathways (Burgess and Shmida, 1985; Downtown, 1975; Ellenberg, 1981; Hesla et al., 1982; Krenzer et al., 1975; Pearcy, 1983; Pearcy and Calkin, 1983; Salisbury and Ross, 1978, 1992; Waller and Lewis, 1979; Winter and Troughton, 1978). In addition, aquatic plants also possess a CO_2 pump system (Taiz and Zeiger, 1998).

The C4 Plants: Plants with C_4 metabolism grow in hot environments, and the leaf anatomy of C_4 plants is different than that of C_3. There are numerous advantages of growing C_4 plants in dry areas. In addition to possible advantage in growth or yield, C_4 plants have a distinct advantage in water and nutrient use efficiencies and tolerance of high temperatures (Gifford, 1974; Hatch, 1987; Brown, 1994; Ku and Edwards, 1977; Sage and Pearcy, 1987). Further, C_4 grasses can produce new leaves at lower levels of plant N than C_3 grasses (Wilson and Brown, 1983). The combination of higher N use efficiency and production of leaves at lower N levels in the C_4 plants can be an advantage in soils deficient in N (Brown, 1978).

CAM Plants: The CAM mechanism is adapted to desert environments, and occurs in members of at least 18 families including *Cactaceae, Orchidaceae, Bromeliaceae, Liliaceae, Euphorbiaceae,* and *Crassulaceae* (Crassula, Kalanchoe, Sedum) and numerous angiosperms. Examples of CAM plants include cacti, euphorbias, pineapple, agave. The CAM plants have a high water use efficiency (WUE), and lose 50 to 100 g of water for every g of CO_2 gained compared to 250 to 300 g for C_4 plants and 400 to 500 for C_3 plants. The CAM mechanism is similar to the C_4 cycle, but has some notably different features. The stomata of CAM plants open at night and close during the day. There are some facultative plants, which carry on C_3 metabolism under unstressed conditions, and shift to CAM metabolism in response to heat, water, or salt stress. Some examples of C_3, C_4 and CAM plants are listed in Table 3-9.

Appropriate use of biotechnological innovations to develop new germplasm can play a major role in desertification control through biological measures. Soil

Table 3-9. Some examples of plant species with different photosynthetic metabolism.

C_3	C_4	CAM
a) Monocots • Oats (Avena sativa) • Barley (Hordeum vulgaris) • Wheat (Triticum aestivum) • Rice (Oryza sativa) b) Dicots • Sugarbeet (Beta vulgaris) • Mustard (Brassica spp) • Sunflower (Helianthus spp) • Peanut (Arachis hypogea) • Cotton (Gossypium hirsutum)	a) Monocats • Bule grama (Bouteloua gracilis) • Mexican teosinte (Euchlaena mexicana) • Sorghum (Sorghum bicolor) • Sudan grass (Sorghum sudanense) • Corn (Zea mays) • Sugarcane (Saccharum officinalis) b) Dicots • Amaranth (Amaranthus spp) • Australian salt brush (Atriplex semibaccata)	Agave Family • American Agave (Agave americana) • Century Plant (Agave parryi) • Lechuguilla (Agave lechuguilla) Succulent plants • Saguaro Cactus (Cornegiea gigantea) • Barrel Cactus (Ferocactus acanthodes) • Desert Prickly Pear Cactus (Opuntia phaeacantha) • Beavertail Cactus (Opuntia basilaris) • Hedgehog Cactus (Echinocereus triglochidiatus) Yucca plants • Soaptree Yucca (Yucca elata) • Mohave Yucca (Yucca schidigera)

scientists and biologists working together can design plant species that are better adapted to harsh environments of drylands. Appropriate plant species are characterized by deep root system, drought-tolerance, toxicity resistance, and high water and nutrient use efficiencies.

Enhancing Water Use Efficiency (WUE)

Lack of adequate water for plant growth is the principal constraint in dry areas. The precipitation is low and erratic, and a considerable proportion of it is lost by runoff and evaporation. Thus, decreasing water losses by runoff and evaporation is critical to enhancing biomass productivity. Experiments conducted in the west African Sahel (Table 3-10), in northern Africa (Tables 3-11 and 3-12), and elsewhere have demonstrated that adoption of improved technologies (e.g., stone bunds, micro-catchment, appropriate tillage methods, improved crop rotations, and adoption of new species, applications of organic amendments, judicious use of fertilizers and especially that of P) can greatly improve WUE, biomass production, and income. Water harvesting techniques can also improve soil fertility (Niemeijer, 1998). A considerable literature exists with example of such technologies for site-specific situations in different dry areas of the world.

Growing appropriate plant species is critical to enhancing the WUE. Rather than growing common cereals (wheat, barely, sorghum) and legumes (chickpea, clovers), there are a wide range of grain crops and legumes which are adapted to drylands (Table 3-13). Some promising ones are grain amaranth, quinoa and triticale among

Table 3-10. Water harvesting by stone bunds and its effects on biomass and grain yield of Pearl millet in northern Burkina Faso in 1986 (Lamachère and Serpantie, 1991).

Treatment	Biomass yield	Grain yield
	--------------------kg/ha------------------	
Traditional	2520	233
Traditional + stone bunds	3010	406
Traditional + stone bunds + tillage	4640	837

*Both biomass and grain yields could have been greatly enhanced had water harvesting techniques been coupled with judicious use of fertilizers, especially phosphorus.

Table 3-11. Effect of improved management systems on erosion control and crop yield on a Vertisol (12% slope) at Ouzera, Algeria (Arabi and Roose, 1992).

Treatment	Maximum runoff (% of rainfall)	Erosion (Mg/ha/yr)	Yield (Mg/ha)	
			Grain	Straw
Traditional system (wheat)	7-16	0.2-0.3	0.7	0.2
Improved alfalfa pasture	0-9	0.05-0.3	—	2.2
Improved wheat-legume rotation	1-8	0.1-0.2	6.5	3.1 (5 beans)
Bare fallow	7-86	2.7-6.0	0	0

Table 3-12. Effects of improved management on runoff control in Ouzera, Algiers (Roose, 1996)

Management	Runoff (% of rainfall)	Income (US $/ha)
Agropastoral system		
Traditional	16	89
Improved	8	1,279
Sylvopastoral (brown soil)		
Degraded	25	---
Reforested	3	---
Grassed	7	---
Orchard (red ferra litic soil)		
Traditional	12	357
Improved	9	1,507
Vineyard (brown colluvial soil)		
Traditional	8.3	1,226
Improved	2.7	2,334

Improved systems involved appropriate tillage methods and recommended cropping systems.

Table 3-13. Some cereals and legumes adapted to drylands (adapted from Hinman and Hinman, 1992).

GRAIN CROPS			
Grain Amaranth	A. hydrochondriacus A. cruentus, A. caudatus	Hot and dry regions (C-4) <200 mm rainfall	High quality protein in grains
Quinoa	Chenopodium quinoa	High altitude, low rainfall, freezing temperature, intense radiation	Highly nutritious grains with saponin coating (bitter)
Triticale	Cross between wheat (Triticale spp) and rye (Secale spp)	Tolerant to salinity and alkalinity, drought resistant	Livestock feed
LEGUMES			
Bambara groundnut	Voandzeia subterranea	Drought tolerant, resistant to pests (SSA)	Nuts are good flavored, and well balanced nutrition
Marama bean	Tylosema esculentum	Drought tolerant, perennial (Kalahari desert)	Produces edible tuber, and seeds of high protein content (30-39%)
Tepary bean	Phaseolus acutifolius	Desert legume (southwest USA and Mexico), thrives in arid and hot climates and poor soils. Matures in 70-75 days. Drought avoidance mechanism	High protein content (20-25%) and starch (40%), and high fiber content
Narrow leaf lupin	Lupinus angustifolius	Drought tolerance	Livestock feed

cereals, and bambara groundnut, marama bean, tepary bean and narrow leaf lupin among legumes. In addition to food crops, there are also several promising plants of industrial importance (Table 3-14). These plants, in addition to being efficient in water use, can improve farm income through sale of industrial material (e.g. resins, rubber, oil and fodder). Many xerophytic plants can be grown for production of fruits, fodder and feedstock and fuel wood (Table 3-15).

Table 3-14. Perennials and low-water-use plants of industrial importance (Hinman and Hinman, 1992)

Name	Botanical name	Attributes	Usefulness
Grindelia (gum weed)	Grindelia camporum	Adaptable to stressful environment	Aromatic resins for use in inks, adhesives, varnishes
Rubber Rabbitbrush (gum weed)	Chrysothamnus sp. C. nauseosus C. albicaulis C. hololeucus C. consimilis C. graveolens	Adaptable to drought and aridity, cold winter and hot summers, nutrient poor soils (C 3 plant) Native to western USA	Natural rubber, resins and oils useful for natural pesticides
Guayule	Parthenium argentatum	Semi-desert perennial shrub, not very cold- and salt-tolerant, needs irrigation	Rubber of high quality
Milk weeds	Asclepias erosa A. speciosa	Herbaceous perennial	Latex and rubber, chemical feedstock and fuel, fiber, energy source
Neem	Azadirachta indica	Drought tolerant, tolerant to saline/sodic soils, fast growing	Biocidal properties, oil of industrial use
Jojoba	Simmondsia chinensis	Drought and heat tolerant (up to 50°C), not tolerant to frost	Source of oil and waxes
Vernonia sp.	Vernonia anthelmintica V. galamensis	Short rainy season of 3-4 months	Epoxy fatty acids, oil (Vernolie acid)
Bladderpod	Lesquerella fendleri	Adapted to dry, carbonate-rich soils; tolerant to heat/drought and cold	Hydroxy fatty acids and oils
Mesquite	Prosopis juliflora P. chilensis P. africana P. tamarugo P. cineraria	Leguminous tree with deep root system, anti-erosion and dune stabilization and BNF capacity, drought/heat tolerant. Tolerant to saline soils and water (6,000 ppm)	Firewood, fodder, mulch material
Gum trees	Eucalyptus spp	Drought tolerant, fast growing	Timber, firewood, industrial oil, pharmaceuticals

Table 3-15. Some miscellaneous xerophytic plants of economic importance (adapted from Stiles, 1988; Hinman and Hinman, 1992; Le Houerou, 1975; Patil et al., 1996).

Name	Species	Important characteristics	Usefulness
Prickly pears (cacti)	Opuntia ficus-indica O. streptacantha O. fusicaulis O. robusta O. lindheimeri O. inermis O. phaeacantha Mammillaria microcarpa Nopalea cochenillifera	Special photosynthetic pathway with high CO_2-fixation efficiency (CAM),* high WUE for arid regions	Specialty crop fruits, vegetable crop, and as fodder and forage
Buffalo gourd	Cucurbita foetidissima	Perennial plant adapted to wastelands with little rainfall. Native of western North America	Seeds rich in oil and protein, and large roots rich in edible starch Fruits can be harvested mechanically
Ye-eb	Cordeeauxia edulis	Perennial shrub adapted to hot and dry poor soils	
Acacia	Acacia albida A. amara A. aneura A. ligulata A. cyanophylla A. salicina A. victoriae A. nilotica A. millifera A. Auriculiformis A. ferugenia	Drought tolerant, soil fertility improvement, soil salinity control	Fodder, shrubs, fuel wood, gums, pharmaceuticals
Henna	Lowsonia spp.	Drought tolerant	Shampoos, soaps, cosmetics
Myrrh	Commiphora spp.	Drought tolerant	Incense, pharmaceuticals, perfumes and flavoring, paper
Vetiver	Vetiveria spp	Widely adapted	Aromatic oil, erosion control, ground cover

*CAM = Crassulacean acid metabolism

Irrigation

Supplementary irrigation is an important technology for enhancing biomass production in arid and semi-arid regions. With proper management, irrigation can improve productivity and enhance soil quality. With improper use, however, irrigation can also lead to waterlogging and salinization. Vast areas of irrigated lands in dry areas have been degraded due to rise of water table and buildup of salts in the soil profile. The WUE of existing irrigated lands can be greatly enhanced by the adoption of improved systems (FAO, 1997). Hillel (1997) proposed the HELPFUL irrigation system (high frequency, efficient, low-volume, partial area, farm unit, low cost). Flood irrigation, the common system used throughout the developing world, is wasteful and most inefficient and can lead to severe problems of land degradation.

In addition to improving the efficiency of existing irrigated lands, there is also some potential of expanding irrigated land area in some regions. Hillel (1997) reported that total potential of irrigable land area in sub-Saharan Africa is 39 Mha. Rather than large-scale irrigation schemes (large dams), the emphasis needs to be on small-scale irrigation projects involving small farm unit. In Kano State, Northern Nigeria, Essiet (1990) reported that more adverse changes took place in soils under largescale irrigation than in those under small holder farms. Appropriate smallscale irrigation schemes may involve use of ground water, runoff storage, water harvesting techniques, micro-catchment farming, and other cost-effective and simple watershed management techniques.

Another aspect of irrigation is extremely relevant for the drylands. In addition to salt-affected soils, some water reserves are saline (brackish) with high salt content ranging from 5000 to 40,000 ppm. While common crops cannot tolerate such high salt concentrations, some halophytes can adapt to such conditions. A major opportunity for biotechnology lies in designing new germplasm that is tolerant to high salt concentration in the root zone, and has high net primary productivity in harsh environments. Such engineered germplasm might also be made to biodegrade slowly. Examples of some useful halophytes shown in Table 3-16 indicate vast potential of producing high-grade fodder, forage, oil, and food. Experiments reported by Glenn et al. (1993) show that halophytes irrigated with sea-water can produce biomass yield of 17 to 35 Mg/ha/yr with a net C sequestration rate of 4-8 Mg/ha/yr (Table 3-17). Further, the residence time of C in drylands is much longer and the decomposition rate much slower than that in humid environments (Gifford et al., 1992).

Water harvesting techniques offer another opportunity to increase available water for plant growth, and improve soil conditions. In eastern Sudan, Niemeijer (1998) reported that the indigenous water harvesting system (teras) offers greater production security and minimizes risks of crop failure.

Table 3-16. Some useful halophytes.

Common name	Botanical name	Important characteristics	Usefulness
Pickle weed	Salicornia spp (SOS-7)	Tolerant to saline water (35,000-40,000 ppm). Wilts when irrigated with fresh water. Biomass production capacity of 15-20 Mg/ha/yr	High-grade fodder and oil seed; 30% of seed is high-grade oil. Straw contains 40% salt that after rising can be used as fodder
Salt grass (wild wheat grain)	Distichlis palmeri	Perennial with no gluten in grains. Important for erosion control. Can be grown in intertidal zone with salt content of 40,000 ppm	Edible grains and forage grains rich in bran and fiber
Ny Pa forage	Distichlis spp.	7-10 Mg/ha of dry matter per cutting with 4 or more cuttings per year. Salt tolerant forage with no accumulation of salt. Tolerant to saline water up to 2,000-15,000 ppm of salt.	Forage as good as alfalfa. Protein content in leaves is 12-25%. Reclamation of salt-affected soils
Salt brushes	Atriplex nummularia A. halimus A. canescens A. lentiformis A. semibaccata A. glauca A. vesicaria	Perennials can be irrigated with saline or brackish water, C 4 metabolism, drought tolerant with deep root system. Biomass production potential of 8-11 Mg/ha/yr	Useful forage crops, can be grown and harvested to remove salts
Spirulina	Spirulina geitleri	Algae adapted to saline/alkaline lakes and ponds in arid regions	High protein food and feed

Table 3-17. Mean annual biomass yield and C sequestration rate of sea-water irrigated halophytes at Puerto Penasco, 1990-1992 (modified from Glenn et al., 1993).

Specie	Biomass yield	C sequestration rate*
	-----------------Mg C/ha/yr	--------------------
Batis maritima	34.0	8.2
Atriplex linearis	24.3	6.7
Salicornia bigelovii		
year one	22.4	5.6
year two	17.7	4.3
Suaeda esteroa	17.2	4.3
Sesuvium partulacastrum	16.7	4.2

*Based on net primary production (NPP)

Erosion control

Accelerated soil erosion has severe adverse impact on soil fertility (Leys and McTainsh, 1994). Some traditional erosion control measures can be effective under site-specific conditions (Van Dijk, 1997); widespread adoption of improved practices is essential for desertification control and restoration of degraded soils. The effectiveness of engineering techniques of erosion control and runoff management can be greatly improved through adoption of biological control measures. Vegetal cover (percent of the land area covered by vegetation) is important to controlling erosion by wind and water, gully control and sand dune stabilization (Lal, 1990; Roose, 1996). In addition to water conservation in the root zone, soil fertility improvement is also critical to establishing the protective vegetal cover (Sanchez and Buresh, 1997: Johnston and Syers, 1996). In the southwestern USA, Bedunah and Sosebee (1986) conducted a 3-year study to eradicate mesquite (*Prosopis gladulosa*) and control soil erosion. Eradicating mesquite increased herbaceous cover of Klein grass (*Panicum coloratum*) and other vegetation, reduced soil erosion and improved SOC content and soil quality. However, eradicating mesquite is a major challenge because of drastic changes in the ecosystem. Other important grass species for improving ground cover and enhancing forage quality of rangelands in southwestern USA include black grama (*Bouteloua eriopoda*), blue grama (*B. gracilis*) and tobosa grass (*Hilaria mutica*) (Seitlheko et al., 1993). Establishing shrubs and trees is also an important strategy for desertification control, restoring eroded soils and improving vegetation cover. The strategy is to plant multipurpose trees for shelterbelts or windbreaks (*Neem* or *Acacia* spp.), for reinforcing riverbanks (*Eucalyptus* or *Populus*), for animal fodder (*Leucaena, Acacia, Balbergia*) and for fuel wood. In Argentina, Bravo et al. (1995) observed that strip cropping controlled wind erosion and enhanced soil quality. They compared changes in soil aggregation and SOC content in strips of soil—the one permanently cropped and the other maintained in weeping lovegrass (*Eragrostis curvula*). The pasture strip had 21% higher mean weight diameter (MWD) of aggregates and 29% more SOC content than the cropped strip. The choice of appropriate species for erosion control depends on soil, climate and other ecological factors.

Controlled grazing is also an important strategy for managing vegetal cover and controlling soil erosion. Stocking rate is an important factor in sustainable management (Abel and Blaikie, 1989). Experiments conducted by Wiggs et al. (1994) in the sand-dune surfaces of the southwest Kalahari Desert showed that destruction of the vegetation canopy increased near-surface wind velocity. Denudation by burning, grazing or drought led to a 3-fold increase in dune surface activity. In the west African Sahel, Kessler et al. (1998) observed that low-intensity grazing did not increase vulnerability to erosion. In the Rajasthan desert of India, Kumar and Bhandari (1992, 1993) observed severe decline in vegetal cover in the grazed sand dune areas. Excluding cattle through erecting exclosure resulted in reduced wind erosion and growth of palatable plants. In contrast to the common belief, however, Oba (1992) observed in northwestern Kenya that soil and vegetation degradation occurred even in the absence of grazing, especially when the health of the plant species growing at the time was dependent on grazing-induced perturbations.

Soil fertility management

Soil fertility improvement is an important aspect of soil quality enhancement and C sequestration in soil and biomass. The appropriate formulation and the rates of application of fertilizers may be different for different soils and land use patterns. Because of uncertain rains and other risks, recommended rates of fertilizer applications are usually low for dry areas. Furthermore, where feasible a combination of organic manures and inorganic fertilizer is a better option than inorganic fertilizer alone. Several long-term experiments in dry regions have demonstrated the importance of judicious use of fertilizer, compost, and nutrient management (Traoré and Harris, 1995; Singh and Goma, 1995; Pieri, 1995; Miglierina et al., 1996; Laryea etal., 1995). Diaz et al. (1997) monitored the impact of urban biosolids on soil quality in semi-arid areas of Spain. Plant cover and biomass increased substantially, and the urban biosolids made an effective amendment to regenerate plant cover on degraded soils.

Nambiar (1995) summarized the data from several long-term soil fertility management experiments conducted in India. For sandy soils of Ludhiana, Punjab, the SOC content of unmanured plots increased from 0.20% in 1971 to 0.25% in 1989. With application of nitrogen, phosphorus, potassium, sulfur, and manure, the SOC content increased in the same period from about 0.20% to 0.40%. For a clayey soil at Jabalpur, central India, the SOC content increased from about 0.6% to 1.1% with recommended inputs for soil fertility management. For black cotton soils in Ethiopian Highlands, Wakeel and Astartke (1996) recommend intensification of agricultural practices to minimize risks of land degradation. Jangir et al. (1997) observed that yield of pearl millet (*Pennisetum glaucumu*) in an arid region of Jodhpur, Rajasthan, increased with application of N fertilizer up to 40 Kg N/ha under rainfed conditions and 80 Kg N/ha with supplemental irrigation. In Gurdaspur and Hissar, Mishra et al. (1974) reported that application of manure at the rate of 9-30 Mg/ha/yr caused significant increase in SOC content. Similar observations were made by Ruhal and Singh (1982). Muthuvel et al. (1989) reported that application of manure to a cotton-pearl millet rotation in dryland Vertisol increased SOC content and crop yield. Chaudhary et al. (1981) reported that SOC content increased by 0.033%, 0.042% and 0.143% by applications of 30 Kg P_2O_5, 60 Kg P_2O_5, and 15 Mg/ha of manure to pearl millet-wheat rotation.

The beneficial impact of soil fertility enhancement on SOC content has been observed elsewhere in dry areas prone to desertification. In Niger, Michels et al. (1995) observed that surface application of millet residue mulch at 2 Mg/ha significantly reduced wind erosion and increased SOC content and cation exchange capacity (CEC) of the surface layer. For Kalahari Desert soils of Botswana, Perkins and Thomas (1993) reported that SOC content of soils near a deep well ranged from 4.8 to 7.6% compared with 0.4 to 0.6% about 1000 m away (Table 3-18). Given the input of organic matter, therefore, SOC content can be increased by an order of magnitude even in harsh environments.

Table 3-18. SOC content of soils of three ranches (A, B, C) in the Kalahari Desert, Botswana, in relation to the distance from water wells. (Perkins and Thomas, 1993).

Distance from well	SOC content (%)		
(m)	A	B	C
0	6.4	4.8	5.0
25	2.0	7.6	1.4
50	2.0	2.2	0.4
200	0.4	0.6	0.6
400	0.4	0.6	0.6
800	0.6	0.6	0.6
1500	0.8	0.6	0.4

Animals tend to congregate near the waterwell and add organic material to the soil in its vicinity.

Crop rotations

Crop rotations are also important to soil quality enhancement. In Argentina, Galantini and Rosell (1997) reported that rotations of mixed pasture (5.5 years) and annual crops (4.5 years) maintained 17.3 Mg/ha of SOC compared with 11.2 Mg/ha in continuous cultivation with a wheat-sunflower rotation. In another study, Miglierina et al. (1993, 1996) observed a high SOC content in wheat-grassland and wheat-alfalfa rotations especially with a conservation tillage system. In eastern Bolivia, Barber (1994) observed that subsoiling and incorporation of cover crops in rotation enhanced quality of degraded soils. In Saudi Arabia, Shahin et al. (1998) observed that introducing alfalfa in rotation with wheat grown on a sandy soil decreased salinity with a three-fold increase in SOC content compared with the control. In Maharashtra, India, Lomte et al. (1993) reported that inter-cropping sorghum with legumes and application of manure increased SOC content and aggregation.

Fallowing

Fallowing, taking land out of agricultural production and permitting natural vegetation to grow, is a common practice used for restoring degraded soils. Bush fallowing is most widely practiced in tropical and sub-tropical agriculture (Nye and Greenland, 1962). In northern Nigeria, Abubakar (1996) monitored the impact of duration of fallowing on changes in SOC content. The mean SOC content of the surface soil was 0.94% for 2-year fallow, 1.13% for 5-year fallow, 1.42% for 10-year fallow and 1.44% for 15-year fallow. There were also increases in SOC content of the sub-soil. The SOC content reached 95% of the equilibrium level within 10 years of fallowing. The data from a 30-year fallowing experiment conducted in eastern Spain showed that SOC content stayed more or less constant at a low level for several years. Subsequently, there was a significant increase in SOC content in the top 10-cm layer and a slight increase in 20-30 cm depth. The data showed that a notable increase in SOC content occurred after 20 years of fallowing. Although natural regeneration can enhance soil quality (Ruecker et al., 1998), fallowing efficiency can be vastly improved by growing appropriate cover crops (Barber and Navarro, 1994).

Residue mulch

Application of biomass (crop residue, tree leaves and branches etc.) to degraded soils can enhance activity of soil fauna and improve soil quality. Termite activity is important to improving soil structure in arid and semi-arid regions. In Burkina Faso, Mando (1997) monitored the effectiveness of mulch and termite activity in the rehabilitation of soil structure. Termite activity improved structure, soil-water regime, biomass production and SOC content. Kessler et al. (1998) also observed that spreading mulch of dead wood, and alignment of dead wood and stones along the contour enhanced regeneration of the vegetation. In Niger, Sterk et al. (1996) observed that application of *Zai* (using pits filled with compost) and mulching with crop residues or tree branches decreased risks of wind erosion and improved productivity. Residue mulches are effective against rill/inter-rill erosion on gentle slopes but not against mass movement on steep slopes. On Vertisols in Karnataka, India, Badanur et al. (1990) reported that application of sorghum stubbles and safflower stalks improved soil structure and increased SOC content. In degraded red soils of south China, Li et al. (1994) observed that applications of organic material to degraded soils enhanced SOC content.

Soil salinity control

There are about 930 Mha of salt-affected soils in the world (Sumner et al., 1998; Sumner and Naidu, 1998). Therefore, reclamation of salt-affected soils is an important aspect of soil quality enhancement and increasing C sequestration in below-ground and above-ground biomass and in the SOC pool. There are numerous proven technologies of reclaiming salt-affected soils (Gupta and Abrol, 1990). Experiments conducted at the Central Soil Salinity Research Institute, Karnal, India, have demonstrated the importance of applying gypsum and organic manures and of leaching and growing appropriate plants for salinity control. The strategy is to enhance SOC content, improve soil structure and infiltration rate and replace Na^+ adsorbed on clay minerals with Ca^{2+} and Mg^{2+}. Translocation of carbonates to subsoil with infiltrating irrigation water has an important ameliorative effect.

Growing salt-tolerant (halophytic) plants can improve above- and below-ground biomass production and increase SOC content. Singh (1989) observed that among several fuelwood species evaluated, *Prosopis juliflora* was most adapted to alkaline soils and produced the most biomass. *Sesbania sesban* and *Tamarix dioca* also exhibited good adaptability. Singh et al. (1994) observed that growing salt-tolerant woody species improved soil quality. Important among these species were *Prosopis juliflora, Acacia nilotica, Casuarina equisetifolia, Tamarix articulata, Leptochloa fusca.* Some of these species caused notable increases in SOC content. The data in Table 3-19 show increase in SOC content by the *Prosopis* + grass treatment. In

Table 3-19. Increase in SOC content of an alkali soil due to reclamative effect of *Prosopis juliflora - Leptochloa fusca silvipastoral* system (Singh et al., 1994).

Time (months)	0-15 cm depth		15-30 cm depth	
	Prosopis	Prosopis + grass	Prosopis	Prosopis+ grass
	%			
0	0.18	0.19	0.13	0.12
22	0.20	0.28	0.12	0.16
52	0.30	0.43	0.19	0.21
74	0.43	0.58	0.29	0.36

comparison with the original SOC content of 0.24%, SOC content in 0-15 cm and 15-30 cm depth respectively was 0.85% and 0.55% for *Acacia nilotica*, 0.66% and 0.33% for *Eucalyptus tereticornis*, 0.93% and 0.58% for *Prosopis juliflora*, 0.86% and 0.58% for *Terminalia arjuna*, and 0.62% and 0.47% for *Albizzia labbek*. In addition to fuel wood, there are also suitable fruit trees which are adaptable to highly alkaline soils. For northwestern India, adaptable fruit trees include jamun (*Syzygrum cuminii*), tamarind (*Tamarindus indica*), ber (*Zizyphus mauritiana*), and guava (*Psidium guajava*) (Singh et al., 1997). Several tree species with beneficial and reclamative effects on alkali soils include *Eucalyptus tereticornis, Acacia nilotica, Albizia lebbeck,* and *Terminalia arjuna*. Improvements in SOC content can set in motion restorative trends leading to increase in soil microbial activity and biomass carbon (Ragab et al., 1990). Over an 8-year period in north central India, Garg (1998) observed significant improvements in SOC content of a sodic soil through planting salt-tolerant tree species. The SOC pool in the top 60-cm layer increased with duration after planting trees (Figure 3-3). The rate of increase was low for the first 2 to 4 years, high (exponential) between 4 and 6 years, and stabilized at a low value between 6 and 8 years.

Salt-tolerant trees and forage species have also proven useful in reclaiming sodic soils in Pakistan (Qadir et al., 1996). In addition to improving structural properties, trees also affect salt balance by lowering of water tables leading to a net downward leaching. In southwestern Australia, Farrington and Salama (1996) recommended that revegetation by trees and shrubs be used to control dryland salinity.

Application of manure and gypsum is also important to improvement of soil structure and reclamation of sodic soils. Use of manure and compost is facilitated by integrating livestock with the cropping system (Chaudhary et al., 1981; Haque et al., 1995; Harris, 1995; Pieri, 1995). In Maharashtra, More (1994) reported that application of farm by-products and organic manures improved quality of sodic Vertisols, enhanced SOC content and increased crop yields. Batra et al. (1997) observed the impact of growing karnal grass (*Leptochloa fusca*) and of applying gypsum on the reclamation of an alkaline soil. There were significant improvements in soil quality within 3 years of applying these treatments. Singh et al.

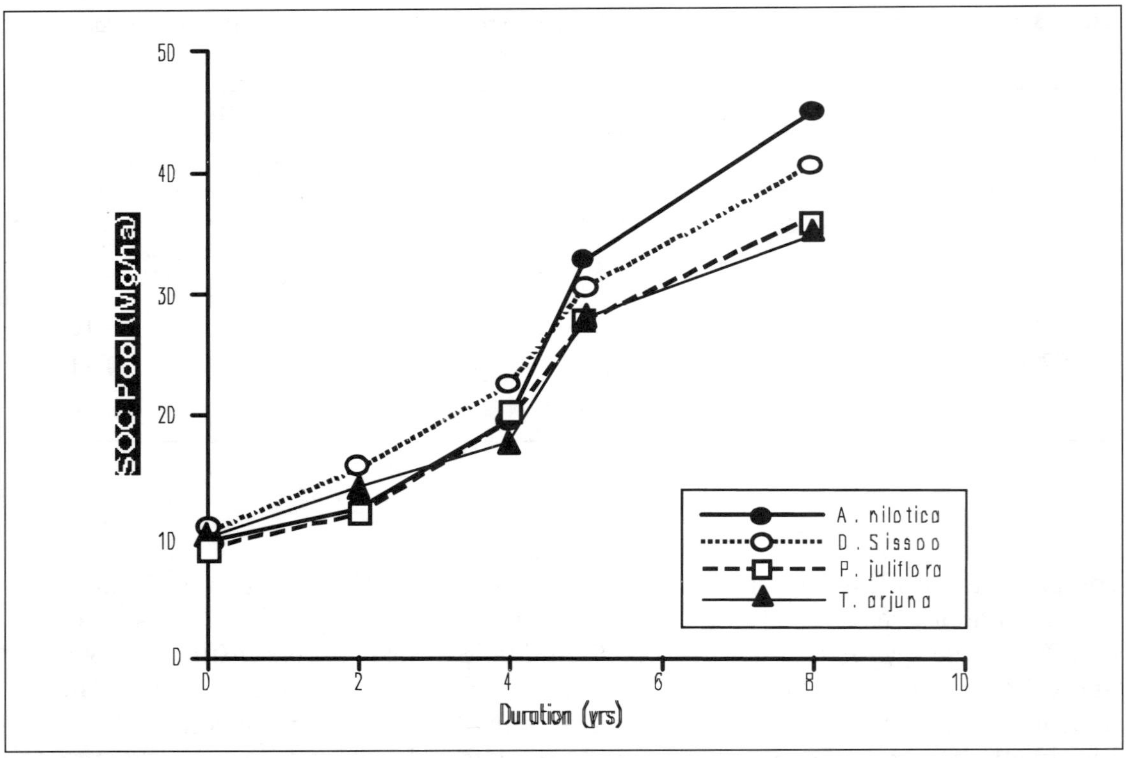

Figure 3-3. C sequestration through reclamation of salt affected soils in northern India (recalculated from Garg, 1998

(1988, 1989) and Singh and Singh (1995, 1996) observed that application of gypsum and farmyard manures enhanced the survival rate of mesquite (*Prosopis juliflora*) on a highly alkaline soil. In southern Australia, Emerson (1995) observed that soil-water retention at low suction increased approximately linearly with SOC content, independently of clay content. Creating and maintaining preferential flow paths are important to leaching soluble salts out of the root zone. Heavily grazed areas are prone to compaction and decline in infiltration rates (Seitlheko et al., 1993). In the Pampa region of Argentina, Dreccer and Lavado (1993) observed that preferrential flow paths of a soil with clayey natric horizon were decreased by the trampling effect of cattle.

FORESTRY MEASURES FOR C SEQUESTRATION IN DRY REGIONS

Widespread deforestation, for fuel wood and other domestic uses, has accentuated the impact of harsh dry environments (Boahene, 1998). Therefore, afforestation is an important strategy. Several multi-purpose trees can grow under harsh environments of drylands, improve soil quality albeit at a slow rate, and sequester C in soil and biomass. Kair (*Capparis decidua*) is one such tree adaptable to drylands of northwest India (Gupta et al., 1989). Growing mesquite (*Prosopis* spp) has been useful in reclaiming salt-affected soils in India. Lack of household energy for cooking is a major constraint among rural communities in dry areas. Deforestation for fuel wood has led to denudation of the landscape and exacerbated risks of erosion and desertification in many places. Establishing fuel wood plantations is critical to desertification control and restoration of degraded soils. Some promising species for fuel wood production, soil quality improvement and desertification control are listed in Table 3-20. In Nigeria, Jaiyeoba (1998) monitored changes in soil properties related to conversion of savannah woodland into pine (*Pinus oocarpa*) and *Eucalyptus camaldulensis* plantations. The SOC content of 0-15 cm depth showed a declining trend during initial stages of tree establishment, then an increase and finally a steady equilibrium value. The equilibrium value was attained in about 16 years. The initial decline was apparently due to the lack of ground cover and low biomass production. That is why growing grass or a cover crop in association with trees is a useful strategy. Garg (1998) monitored changes in properties of a sodic soil under four different tree species in north-central India. Results showed a marked improvement in biogeochemical characteristics in general but SOC content in particular. The SOC pool increased from less than 10 Mg/ha to about 45 Mg/ha (about a 5-fold increase) over an 8-year period. Dry soil bulk density of the 0-15 cm layer of the unplanted site was 1.8 Mg/m^3 compared with that of 1.61 Mg/m^3 under *Acacia nilotica*, 1.50 Mg/m^3 under *Dalbergia sissoo*, 1.43 Mg/m^3 under *Prosopis juliflora*, and 1.55 Mg/m^3 under *Terminalia arjuna*. A list of relevant tree species for restoration of degraded soils, and improving biomass production (for fodder, fuel and other uses) is shown in Table 3-21.

Table 3-20. Tree species suitable for firewood and biofuel production in dry areas (adapted from Le Houerou, 1975; Mainguet, 1991; Singh et al., 1994; Patil et al., 1996).

Common name	Botanical name
Tamarisk	*Tamarix* spp
Gum trees	*Eucalyptus* spp
Leucaena	*Leucaena* spp
Cypress	*Cupressus* spp
Casuarinas	*Casuarina equisetifolia* and other spp.
Mesquite	*Prosopis* spp
Neem	*Azadirachta* sp
Acacia	*Acacia* spp (see Table 3-15)
Teak	*Tectona grandis*
Casia sp.	*C.Siamea*
Other species	*Dalbergia* spp
	Khaya spp
	Albizia spp
	Cassia spp
	Parkia spp
	Terminalia arjuna
	Pongamia pinnata
	Sesbania sesban
	Morus alba
	Populus deltoides

Table 3-21. Influence of tree cover on improving SOC content and enhancing soil quality.

Country	Constraint	Sites	Tree species	Reference
Sudan	Drought, wind erosion	Wakey, Baramio	Acacia tortilis	Alstad and Vetaas (1994)
India	Drought	Rajsthan, Haryana	Kair (*Capparis decidua*)	Gupta et al. (1989)
India	Alkali soils	Haryana	Mesquite (*Prosopis juliflora*)	Singh (1989) Singh and Singh (1995)
India	Alkali soils	Haryana	Eucalyptus tereticornis Acacia nilotica, Albizia lebbeck, Terminalia arjuna	Singh et al. (1992)
India	Saline soils	Karnatka	*Casurina equisetifolia, Acacia auriculiformis, Dalbergia sissoo, Syzygium cumini, Pongamia ponnata, Gliricidia maculata*	Patil et al. (1996)
India	Drought	Karnatka	*Leucaena leucocephale*	Badanur et al. (1990)

POTENTIAL OF EROSION CONTROL MEASURES IN DRY AREAS FOR CARBON SEQUESTRATION

Land area affected by strong and extreme soil erosion is 104 Mha; an additional 424 Mha is moderately eroded soil. The strongly and extremely degraded areas must be taken out of agricultural/pastoral land uses and put under restorative measures. Trees and shrubs can be grown on these lands for preventing soil erosion and for generating biofuel. Planting suitable tree species would have three distinct benefits: (a) provide the much needed ground cover and root system for protecting the soil against further erosion by wind and water, (b) produce biomass that can be used as fuel, and (c) enhance SOC content and sequester C in soil. Suitable species have deep root systems to anchor the soil and protect the seedlings against drought stress.

Once established (2 to 3 years after planting/sowing), the above-ground biomass production potential in these environments is 2 to 3 Mg/ha/yr with a total production of 208 to 312 million Mg/yr possible. With a fuel efficiency of 0.7 for direct use (relative to 1 for fossil fuel), the biofuel C offset for these lands is 144 to 216 x 10^6 Mg C/yr (0.14 to 0.21 Pg C/yr). In addition, some of the below-ground biomass produced will be converted to humus or SOC. Assuming a low rate of increase of SOC content at 40-60 kg C/ha/yr, the potential of C sequestration in soil is 4 to 6 x 10^6 Mg C/yr (0.004 to 0.006 Pg C/yr).

There are 424 Mha of moderately eroded lands in dry areas. Recommended agricultural management practices need to be adopted on these lands for soil erosion control and soil quality enhancement. Such practices include installation of engineering devices (e.g., diversion channels, drop structures, infiltration ditches, stone filters or gabiens, terraces) and adoption of appropriate management systems for soil (tillage methods, soil fertility management, residue management, soil amendments) and crops (e.g., rotations, agroforestry techniques, integrated pest management). Adoption of improved practices would control soil erosion, improve crop yield, enhance soil quality and sequester C due to increased SOC content. Enhancing SOC content in dry climates is a challenging task, and the rate of increase in SOC may be low even under the ideal conditions. Assuming the potential rate of increase in SOC content at 80 to 120 kg C/ha/yr, the total potential of C sequestration in moderately eroded soils is 34 to 51 x 10^6 Mg C/yr (0.03 to 0.05 Pg C/yr).

The total potential of C sequestration through soil erosion management is shown in Table 3-22. Adoption of erosion control measures and recommended management practices would decrease C emissions from erosion-displaced sediments by as much as 50 to 75% of the estimated emission (0.13-0.20 Pg) (Lal, 1995). Allowing strong and extremely eroded lands to revert to restorative plantation can lead to fossil fuel offset at 0.14 to 0.21 Pg C/yr, and SOC sequestration of 0.004 to 0.006 Pg C/yr. Adoption of recommended agricultural practices on slightly and moderately eroded lands (Oldeman, 1994) has the potential to sequester 0.06 to 0.10 Pg C/yr. Thus the total potential of C sequestration through soil erosion management in dry areas is about 0.33 to 0.52 Pg C/yr (Table 3-22).

Table 3-22. Potential of C sequestration through restoration of degraded soils.

Technological options	C sequestration potential (Pg C/yr)
1. Restoration of eroded soils	
(a) Erosion control (50-75% of emission reduction)*	0.13-0.20
(b) Restoration of strongly/extremely eroded soils	0.004-0.006
(c) Adoption of recommended practices on slightly/ moderately eroded soils	0.06-0.10
2. Fossil fuel off-set	0.14-0.21
Total	0.33-0.52

*Adoption of conservation-effective measures would lead to reduction in emissions of carbon dioxide from erosion-cased displacement of soil carbon and its enhanced mineralization.

AGRICULTURAL INTENSIFICATION FOR SOC SEQUESTRATION

Adoption of improved/recommended practices on agricultural and pastoral lands is an important strategy for agricultural intensification. There are numerous recommended practices for sustainable use of soil and water resources that have proven successful under on-farm conditions and are awaiting adoption by farmers in dry areas (see the discussion earlier in this chapter). Recommended practices include those for soil-water conservation and management, irrigation management, soil fertility enhancement (including inorganic fertilizers and organic amendments), residue management and tillage methods, improved varieties and associated cropping systems, and integrated pest management (including effective weed control measures).

Programs of agricultural intensification can be applied to prime agricultural land, land that has no or slight degradation. The land area under no or slight degradation is estimated at 427.3 Mha. Because of high inherent soil quality and favorable biomass production, these soils can undergo agricultural intensification through better biomass production. Further, plants with C-3 photosynthetic pathway generally have increased C fixation rates when CO_2 levels in the atmosphere are increased. Enhanced WUE due to partial stomatal closure in CO_2-enriched environments may also lead to a higher biomass production (Ojima et al., 1993). Assuming that adoption of improved land use and farming/cropping systems may lead to increase in SOC content in these soils at the rate of 30 to 50 kg/ha/yr, the total potential of C sequestration is 12.8 to 21.4 x 10^6 Mg C/yr (0.013 to 0.021 Pg C/yr).

OPTIONS FOR C SEQUESTRATION BY DESERTIFICATION CONTROL MEASURES

Improving soil quality is the principal strategy for C sequestration. Increasing biomass production, decreasing losses (of soil, water and nutrients) out of the ecosystem, enhancing soil biodiversity (activity and species diversity of soil flora and fauna), and strengthening of nutrient cycling mechanisms improve soil quality, increase aggregation, and enrich SOC content in the stable fraction with relatively longer turnover time. Desertification control is an important global strategy for C sequestration in soils (Squires et al., 1995). There are two principal strategies: (a) restoring severely degraded soils, and (b) adopting recommended land use and soil/water/crop/vegetation management technologies.

Soil erosion: Soil erosion, a principal form of degradative process, has been discussed above.

Physical and chemical degradation: In addition to erosion, drylands also are prone to physical and chemical forms of soil degradation. Soil physical degradation involves decline in soil structure leading to crusting, compaction, hard-setting and exposure of plinthite that eventually hardens into a laterite (Lal et al., 1989). Soil chemical degradation includes salinization, fertility depletion or nutrient imbalance, and acidification. Land area affected by strong and extreme forms of physical and chemical degradative processes is 34 Mha (UNEP, 1991). Similar to severely eroded lands, these lands may also be taken out of production and restored through planting appropriate trees and shrubs. In addition to production of biofuel (see the following section), soil restoration would also improve SOC content, albeit at a low rate of about 40 to 60 kg/ha/yr. Therefore, the potential of SOC sequestration in these soils is 1.4 to 2.0 x 10^6 Mg C/yr (0.001 to 0.002 Pg/yr).

In addition, there are 46 Mha of moderately degraded soils. Assuming that adoption of recommended practices on these soils may lead to C sequestration at the rate of 80 to 120 kg/ha/yr, the potential of SOC sequestration on these lands is 3.7 to 5.6 x 10^6 Mg C/yr (0.004 to 0.006 Pg C/yr). Therefore, the total potential of C sequestration through restoration of physically and chemically degraded soils is 0.005 to 0.008 Pg C/yr.

SOC sequestration through reclamation of salt-affected soils: There are 930 Mha of salt-affected soils in arid and semi-arid regions of the world. Adoption of reclamative measures on these soils can lead to increases in above- and below-ground biomass production and to improvements in SOC content. Assuming the rate of SOC increase through adoption of reclamative measures is 200 to 400 kg C/ha/yr (the rate was 3 to 4 Mg C/ha/yr in some soils in north central India), the potential for C sequestration is 0.186 to 0.372 Pg C/yr.

Calculations made above are tentative due to possibilities of double accounting. Soils prone to physical and chemical degradation are also susceptible to accelerated soil ersoion. However, the database is not precise enough to assess areas specifically affected by each soil degradative process.

BIOFUEL OFFSET THROUGH DESERTIFICATION CONTROL

Strongly and extremely degraded soils should be taken out of agricultural and pastoral land uses and planted to specific trees, shrubs or grass species that can be used as biofuel. The biomass can be used as direct fuel for power generation. Total land area of strongly and extremely degraded lands (erosion, physical and chemical degradation) in dry areas is 138 Mha. Growing appropriate biofuel crops on such lands is an important strategy. Once established (2 to 3 years after planting/sowing), the above-ground biomass production potential in these environments is 2 to 3 Mg/ha/yr with a total production of 275 to 413 x 10^6 Mg C/yr. With a fuel efficiency of 0.7 for direct use, the biofuel C off-set for these lands is 192 to 289 x 10^6 Mg C/yr (0.19 to 0.29 Pg C/yr).

CARBON SEQUESTRATION POTENTIAL AS SECONDARY CARBONATES IN DRYLANDS

The role of soil inorganic C (SIC) in C sequestration is not well understood. The SIC may be a sink, source or have no effect upon C sequestration depending on site-specific conditions. In systems of partial or complete soil leaching, the major mechanism for sequestered SIC is via movement of HCO_3^- into groundwaters or closed systems isolated from the ambient environment. Reconstructions of carbonate fluxes in soils formed in strongly calcareous parent materials over geological time periods allows calculation of upwards of one Mg of SIC/ha/yr. While this mechanism may appear to be of limited impact for degraded drylands, certainly it has implications when groundwaters undersaturated with $Ca(HCO_3)_2$ are used for irrigation. Enhanced biomass primary productivity and salinity control strategies (e.g., gypsum amendments, organic wastes) can result in increased leaching of

$Ca(HCO_3)_2$ via irrigation if the irrigation waters are not already saturated with respect to bicarbonates.

Dissolution of exposed carbonates in soil systems by acid rain, nitrogenous fertilizers, oxidation of iron sulfides, or organic acids may result in a release of CO_2 unless subsequent precipitation of pedogenic carbonates occurs. However, if the dissolved bicarbonates are either leached through the soil or removed by overland flow without subsequent precipitation of carbonates, then this mechanism is a transient sink of intermediate to long residency rather than source of atmospheric CO_2. The rate of C sequestration through formation of secondary carbonates is a debatable issue. Some researchers argue that the rate is slow (3-5 $g/m^2/yr$) and of little significance (Schlesinger, 1997). Others, however, support the idea that rate of sequestration of atmospheric C may be much higher (personal communication L.P. Wilding, Texas A&M Univ., 1998), with a maximum rate of 114 to 124 kg C/ha/yr. Formation of secondary carbonates is accentuated by biotic activity (e.g., root growth, termites) (Monger and Gallegos, 1999).

The data in Table 3-23 show the potential of C sequestration through formation of secondary carbonates with a range of 0.008 to .426 Pg C/yr. The wide range is indicative of the high variation and large uncertainty due to differences in soil profile characteristics, moisture and temperature regimes, land uses and ecoregional characteristics.

Table 3-23. Estimates of C sequestration through formation of secondary carbonates.

Ecoregion	Land area (10^6 km^2)	Potential rate of C Sequestration* (kg/ha/yr)	Total sequestration Potential (Pg C/yr)
Arid	25.5	0-1	0-0.0026
Semi-arid	23.1	3-114	0.0069-0.2633
Sub-humid	13.0	1-124	0.0013-0.1599
Total			0.0082-0.4258

*Potential rates of C sequestration made available by the courtesy of Dr. L.P. Wilding of Texas A&M University, College Station, TX.

POTENTIAL OF DESERTIFICATION CONTROL TO SEQUESTER C AND MITIGATE THE GREENHOUSE EFFECT

The total potential of restoration of degraded lands and desertification control to sequester C is shown in Table 3-24. The total potential is 0.9 to 1.9 Pg C/yr, with a mean of about 1.4 Pg C/yr. The relative contribution of different components is outlined in Figure 3-4. The options with high C sequestration potential include erosion control (36%), biofuel C offset (29%), reclamation of salt-affected soils (21%), and the deposition of secondary carbonates (14%). Restoration of eroded and salt-affected soils and erosion control are important strategies. Squire et al. (1995) estimated that management of drylands through desertification control has the C sequestration potential of 1.0 Pg C/yr. These high estimates are in contrast to the overall low C storage potential of world soils estimated by Schlesinger (1990). Estimates presented in Table 3-24 are crude, tentative, and merely suggestive of the potential that exists in adoption of judicious land use measures in drylands. Uncertainties are substantial-perhaps 30 to 50%. Further, estimates of potential for different strategies are not additive, and the data need to be judged with due consideration of site-specific conditions.

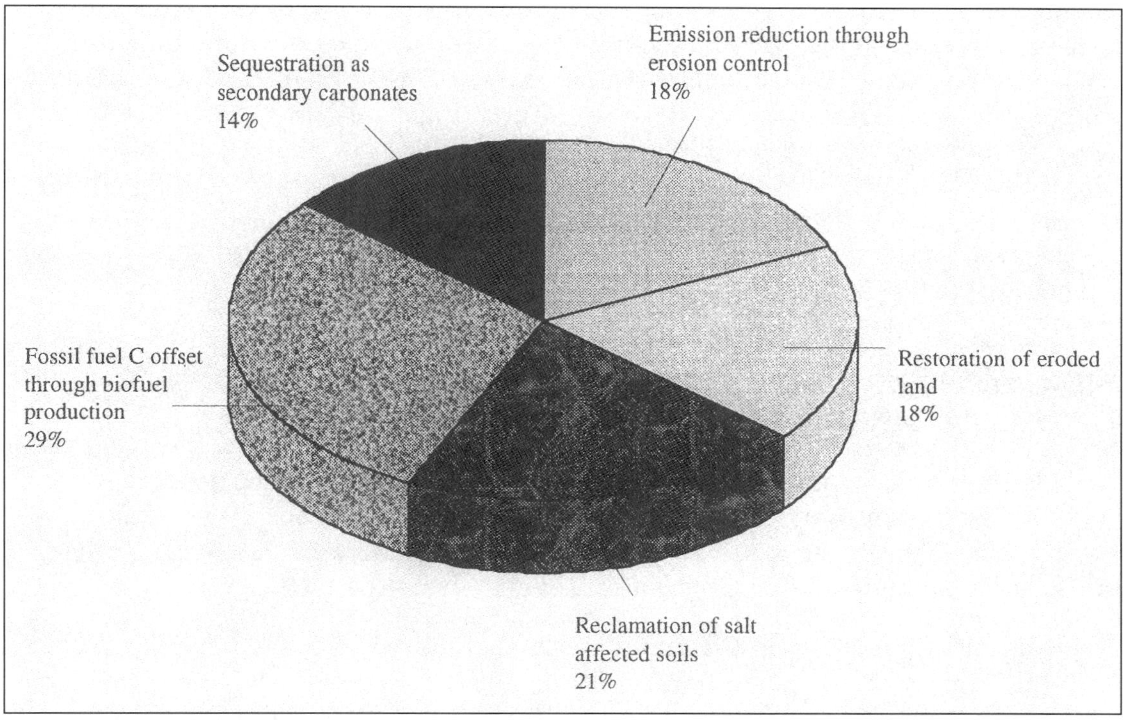

Figure 3-4. Relative contributions of desertification control measures to sequester C and mitigate the greenhouse effect.

Table 3-24. Potential of desertification control and land restoration to sequester C (Pg C/yr)

Process	Range	Mean	% of total potential
Emission reduction through erosion control	0.2-0.3	0.25	18
Restoration of eroded lands	0.2-0.3	0.25	18
Restoration of physically and chemically degraded soils	<0.01	<0.01	---
Reclamation of salt-affected soils	0.2-0.4	0.3	21
Agricultural intensification on undegraded soils	0.01-0.02	0.015	---
Fossil fuel C offset through biofuel production	0.3-0.5	0.4	29
Sequestration as secondary carbonates	0.01-0.4	0.2	14
Total	0.9-1.9	1.4	100

These estimates have high uncertainties, the potentials of different strategies may not be additive, and adoption of recommended measures at global scale is a major challenge to humanity.

ECONOMIC FEASIBILITY OF C SEQUESTRATION IN DRYLANDS

Important issues of economic analyses need to be addressed. Farmers adopt new technology when potential to profit from adoption is high. Policy makers need to provide farmers with new economic incentives to improve production without jeopardizing the natural resource base. A carbon credit or subsidy may be an additional source of revenue to overcome economic constraints. Site-specific strategies must address the economic status of the landowner and the country, and population growth. Integration of livestock with crops may be important to land restoration in some ecoregions, so enhancing income through mixed farming may be crucial to increasing investment in land resources. The issue of C sequestration in soil may be appropriately linked to other interacting concerns (e.g., salinization, erosion, and water quality). Finally, economic incentives to farmers can be determined on the basis of the cost of a unit amount of soil C. The cost-curve analysis requires assessment of the cost of C sequestration in relation to the amount of C sequestered or the land area specifically managed for sequestration (Ravindranath and Somashekhar, 1995; Marland et al., this volume).

INDICATORS OF LAND QUALITY TO MONITOR TRENDS IN CARBON SEQUESTRATION

Land is the sine qua non for agricultural production and food security, but it is also the key to improved environmental management. Land acts as both source and sink for greenhouse gases, recycles nutrients, ameliorates and filters pollutants, and transmits and purifies water as part of the hydrologic cycle.

Maintaining and enhancing the quality of land resources is a key requirement for sustainable rural development (World Bank, 1997), and developing indicators by which to monitor change in land quality is fundamental. However, land quality, like the concept of sustainable land management of which it is a part, requires operational definitions and specific, measurable indicators if it is to be more than an attractive, conceptual phrase. Indicators are already in regular use to support decision making at global, national, and subnational levels for air and water quality, but few such indicators are available to assess, monitor, and evaluate changes in the quality of land resources. Land refers not just to soil but to the combined terrain, water, soil, and biotic resources that provide the basis for land use. Land quality refers to the condition, or "health," of land, specifically to its capacity for sustainable land use and environmental management. Land quality indicators (LQIs) address the dual objectives of environmental monitoring and sector performance monitoring for managed ecosystems (agriculture, forestry, conservation, and environmental management) (Pieri, 1995).

Although a single indicator of land quality is not realistic, a very large number reflecting all possibilities is also not useful. To help resolve this problem, a core set of strategic land quality indicators in use include nutrient balance, yield gap, land use intensity, land use diversity (agrodiversity), and land cover. LQIs being researched include soil quality, land degradation, and agrobiodiversity. (See Appendix I for more detailed descriptions.)

These strategic indicators are the biophysical components of sustainable land management. Although useful in their own right, they must still be complemented with indicators of the other pillars of sustainable land management: economic viability, system resilience, and social equity and acceptability.

The issue of sequestering carbon in the soil must be viewed within the broader context of land quality and sustainable land management. Certain of the LQIs have direct association with soil carbon sequestration, but others link indirectly. For example, the core LQIs dealing with land cover, soil quality, land degradation, and agrobiodiversity link directly because of the emphasis on monitoring soil organic matter and the important roles of soil biology. On the other hand, indicators such as nutrient balance, land use intensity, land use diversity, and yield gap are indirectly linked because they incorporate dimensions beyond organic matter.

An important conjuncture between land quality and soil carbon sequestration is that both focus on the use of tested and well calibrated models, such as EPIC and CENTURY, coupled with remote sensing and selected field sampling, to predict the impacts of improved land management. A field monitoring program depending on site measurements to prove carbon fluxes to the satisfaction of the international community in relation to commitments made under the Kyoto Agreement would be a very high cost greenhouse gas mitigation strategy. Building and verifying reliable carbon content prediction procedures is considerably more cost effective.

Carbon sequestration through better management of agricultural soils is not a panacea, but there is growing evidence that its potential is too large to be ignored (A. Donnelly, personal communication, GEMCo[1]). Particularly important are technologies such as reduced and zero tillage, continuous and rotational cropping, and improved forage and rangeland management. But other techniques are also promising. Consequently, significantly more attention must be given to sequestering carbon in agricultural soils for the following reasons:

- Unsustainable agricultural practices are among the key drivers of deforestation and uncontrolled forest fires in many of the developing nations. It is unlikely that reforestation, afforestation, or forest preservation programs will succeed in the long run unless they are part and parcel of strategies to introduce sustainable land management practices in agricultural areas. Evidence from Africa (P. Sanchez, personal communication) indicates that there are

---

[1]GEMCo is made up of leading Canadian energy and chemical manufacturing companies that are committed to take action on the issue of climate change. Each organization has implemented its own corporate greenhouse gas action strategy, but at the same time GEMCo members have identified the importance of an integrated and cooperative approach in research, planning and project implementation to effectively address the issue of global climate change. GEMCo members directly employ over 60,000 Canadians and account for roughly 25% of Canada's anthropogenic emissions of greenhouse gases. Members are actively participating in domestic and international greenhouse gas offsets. GEMCo is currently co-funding (almost 50%) a large project to fully identify how much incremental carbon can be stored in agricultural soils in Canada. Preliminary estimates indicate that the potential is at least 14 million tonnes of CO_2 each year for the next 20 years, which is approximately 8.5% of the annual greenhouse gas reductions required to meet Canada's commitment under the Kyoto agreement.

four times more trees on-farm in Kenya than in plantations, and that woody biomass in trees outside forest areas is two times greater than the combined biomass in natural forests and tree plantations.

- Planting trees while continuing to farm in an unsustainable manner is not likely to generate real net carbon gains in the long run, because it will not be possible to plant trees fast enough to keep up with deforestation caused by demand for agricultural land in many key regions. Focusing primarily on tree planting is focusing on the symptoms, not on the disease.
- There need not be a trade-off between building up organic carbon in the soil and producing above-ground biomass: sustainable land management practices can lead to increases in carbon stored below the ground coincident with increases in production of above-ground biomass. There are few other such "least cost" emissions mitigation methods available, and few that are more appropriate.
- Programs to promote soil carbon sequestration are typical "win-win" situations. While additional carbon is sequestered from the atmosphere, this may also increase biomass production above ground, improve soil moisture storage, reduce erosion, reduce fossil fuel consumption on the farm, reduce the need for fire to manage farmland, and reduce farmers' demand for and use of certain herbicides. While use of some nitrogen-based fertilizers might increase, currently available data suggest that incremental increases in N_2O emissions (in CO_2 equivalents) are not significant in comparison to incremental decreases in CO_2 emissions from sustainably managed soils.
- In many regions, shifting to more intensive land management through use of sustainable land management practices can produce a positive net financial return to farmers on a 7 to 10 year payback basis. Unfortunately, even in developed countries, few farmers have sufficient capital to consider an investment with a 7-year payback, considering that the first 5 years pose the highest risk for the farmer in terms of investment in farm equipment and possible reductions in yield during the transition period. In these instances, linking with private and public sector institutions to absorb the initial transaction costs through programs like "carbon offsets" provides an enabling environment to initiate adoption of these mitigating measures. This will require, among other things, that soil carbon credits become recognized, tradable commodities under the Kyoto Protocol.

DATA UNCERTAINTIES AND RELIABILITY

The extent of desertification is a debatable issue because of data uncertainties and unreliability, especially with regard to the statistics from developing countries (Carpenter, 1989; Blaikie, 1985; Blaikie and Brookfield, 1987). The quality of data is questionable, and the data are often obtained by unsatisfactory and unstandardized methods. Consequently, errors in data are common. The problem is especially severe in the estimates of soil erosion rates, which often differ by orders of magnitude (Lal, 1994a). Available statistics on land degradation are severely constrained by the lack of experimental data on the rates of soil formation; the on-site impacts of erosion on crop growth and yield; and off-site impacts on water quality, air quality, greenhouse effects and civic structures. (Lal, 1994b; Blaikie, 1989; Blaikie and Brookfield, 1987). Blaikie (1985) observed that available statistics are seldom in the right form, are hard to come by, harder to believe, and difficult to interpret. There is an urgent need to improve methods of data collection and analysis, develop and standardize procedures for conducting experiments and recording observations on appropriate parameters, and identify indicators of land degradation and desertification. The problem is exacerbated by at least three important factors:

- *Scaling up*: In addition to the lack of standard methods, there are also problems with scaling and extrapolation. Soil degradation processes are often site-specific and monitored at plot, soil or landscape scale. The extrapolation of such location-specific data to watershed, ecoregion, state or national scale can lead to gross errors of several orders of magnitude. Some soil-specific processes cannot be extrapolated to watershed or ecoregional scales because of a diverse conglomerate of numerous soils and land uses.

- *Impact*: Land use change inevitably alters soil and vegetation characteristics, and it is difficult to assess the threshold level at which the change becomes degradation. The threshold level depends on critical limits of key land attributes and soil characteristics, and of the impact of land use and management on productivity and ecosystem services. The lack of data on impact leads to the perception of a high rate of improvement with adoption of recommended practices.

- *Resilience:* The severity of desertification and land degradation also depends on land resilience or its ability to restore itself. The problem of desertification is indeed severe for systems that lack resilience. Strongly resilient land and ecosystems are able to recover when the cause of perturbation is removed. In such a situation, the problem is not severe even if the apparent losses are high. The ecosystem's resilience depends on many exogenous and endogenous factors, and management plays an important role in it.

RESEARCH PRIORITIES

Despite these uncertainties the authors believe that the potential for C sequestration is significant—perhaps as much as 1.4 Pg C/yr. The potential is, therefore, more than 40% of the total amount of annual atmospheric increase in CO_2 concentration. If the potential of soil C sequestration were to be realized, it would make a major contribution toward mitigating greenhouse warming. However, estimates of C sequestration potential are tentative because of several constraints, including (a) scarcity of reliable data on soil properties and processes, (b) scaling up data without using the soil or pedon as the basis of extrapolation, (c) difficulties of avoiding double accounting, and (d) lack of analyses to assess the economic feasibility of C sequestration.

There is still uncertainty on how much carbon can be stored in agricultural soils even if improved, sustainable land management can restore soil carbon to pre-cultivation levels. Currently, the Kyoto Agreement does not include soil carbon sequestration, but it may do so in the future. This may require that targets for soil carbon sequestration be set, and monitoring procedures for evaluation and compliance, including land quality indicators, will have to be developed.

Soil carbon sequestration is not a panacea, and there is increasing evidence that a package of greenhouse gas mitigation measures will be necessary to meet the targets of the Kyoto Agreement. However, the amount of carbon that could be stored in the soil is too large to ignore. Further, increasing carbon in the soil is one of the few truly "win-win" strategies with direct benefit to farmers as well as global environmental management. Although the strategy can be implemented in all regions, and it is thus highly suited for carbon offsets, the potential varies considerably according to local soils and the management practices used, and indicators of land quality will be necessary to track performance.

There are no standard indicators to evaluate or predict desertification, especially for countries in temperate regions (e.g., Iceland). The potential of using remote sensing techniques needs to be explored (Mohammed et al., 1996; Dijk, 1997). Use of remote sensing techniques may involve study of spectral behavior of eroded lands, including gullied lands (Kumar et al., 1997), salt-affected soils with high albedo because of salt incrustation (Rao et al., 1998), and lands vulnerable to other degradative processes. Land vulnerability to water erosion can be assessed on a grid of 1/2 by 1/2 (Batjes, 1996). Future research on desertification control should consider the following strategies, objectives and research needs.

Objectives

Principal objectives of adopting appropriate land use and improved technologies are (a) Conserve soil and water; (b) improve soil chemical, physical and biological quality; (c) strengthen nutrient cycling mechanisms, and create a favorable elemental and water balance; (d) enhance vegetal cover and biomass productivity; (e) improve economic returns from degraded soils; and (f) enhance effective rooting depth.

Strategies

A principal strategy is to restore degraded soils and ecosystems through the following: (a) controlling desertification and restoring soils degraded by different processes to enhance biomass productivity and improve soil quality; (b) delineating some bright (trigger) spots in semi-arid and sub-humid ecoregions with high potential for C sequestration (hyper-arid and arid regions have low potential for C sequestration); (c) identifying and cultivating appropriate (economic/industrial) halophytic plants, including irrigation with saline water that can lead to increased productivity and C sequestration in desert environments with salt-affected soils and brackish water; (d) adapting and growing C_4 and CAM plants with high water and nutrient use efficiencies to improve biomass production in harsh environments and lead to C sequestration; (e) using biotechnology for developing new germplasm tolerant to saline and other environmental stresses; and (f) applying organic by-products to land and adoption of techniques that may lead to formation of secondary carbonates in arid regions.

Research and development needs

There are numerous uncertainties and knowledge gaps that require additional research. Principal research needs include—

- Quantifying the extent and severity of soil degradation by different processes (erosion, salinization, pollution/contamination, physical degradation, chemical degradation) at regional, national and global scales
- Evaluating soil C (organic and inorganic) pools and fluxes as influenced by different soil degradative processes
- Identifying practices of desertification control and restoration of degraded soils
- Assessing extent and severity of biological degradation
- Identifying local, regional, national and global policies that will lead to adoption of soil restorative measures
- Evaluating the economic and societal value of unit quantity of C in soil, and develop mechanisms for trading C credits
- Assessing the potential of C sequestration in soil and biomass through restoration of soil degraded by different processes
- Developing means to identify early warning symptoms of desertification and land quality degradation via biological indicators
- Quantifying the role of SIC in C-sequestration processes, rates, dynamics and magnitude
- Using biotechnology for developing plant species that adapt to physical and chemical degraded soil conditions, enhance C-sequestration, and provide for food, fiber and fuel supplies
- Enhancing efficiency of irrigation systems via precision agriculture technology to adjust application of dosing rates to fit variable physical and chemical conditions of soil system
- Conducting economic feasibility studies of different desertification control measures for C sequestration.

ACKNOWLEDGEMENTS

The authors are grateful to Drs. Norm Rosenberg, Larry Wilding, Pierre Crosson and Robert Jackson for their critical comments and review of the manuscript.

REFERENCES

Abel, N.O.J. and P.M. Blaikie. 1989. Land degradation, stocking rates and conservation policies in the communal rangelands of Botswana and Zimbabwe. *Land Degrad. & Rehab.* 1:101-124.

Abubakar, S.M. 1996. Rehabilitation of degraded lands by means of fallowing in a semi-arid area of northern Nigeria. *Land Degrad. & Develop.* 7:133-144.

Alstad, G. and O.R. Vetaas. 1994. The influence of *Acacia tortilis* stands on soil properties in arid northeastern Sudan. *Acta-Oecologica* 15:449-460.

Arabi, M. and E. Roose. 1989. Influence de quatre systèmes de production en région méditerranéenne de moyenne montagne algérienne. *Bull. Réseau Erosion* 9:39-51 (ORSTOM, Montpellier).

Auberville, A. 1949. *Climats, forêts et désertification de l´ Afrique tropicale.* Société d´ Editions Geographiques et Coloniales, Paris.

Badanur, V.P., C.M. Poleshi and B.K. Naik. 1990. Effect of organic matter on crop yield and physical and chemical properties of a Vertisol. *J. Indian Soc. Soil Sci.* 38:426-429.

Balpande, S.S., S.B. Deshpande and D.K. Pal. 1996. Factors and processes of soil degradation in Vertisols of the Purna Valley, Maharashtra, India. *Land Degrad. & Develop.* 7:313-324.

Barber, R.G. 1994. Soil degradation in the tropical lowlands of Santa Cruz, eastern Bolivia. *Land Degrad. & Rehab.* 6:95-108.

Barber, R.G. and F. Navarro. 1994. The rehabilitation of degraded soils in eastern Bolivia by sub-soiling and the incorporation of cover crops. *Land Degrad. & Rehab.* 5:247-260.

Barrow, C.J. 1991. *Land Degradation.* Cambridge University Press, Cambridge, 295pp.

Batjes, N.H. 1996. Global assessment of land vulnerability to water erosion on a 1/2 by 1/2 grid. *Land Degrad. & Develop.* 7:353-366.

Batra, L., A. Kumar, M.C. Manna and R. Chhabra. 1997. Microbiological and chemical amelioration of alkaline soil by growing karnal grass and gypsum application. *Expl. Agric.* 33:389-397.

Beaumont, P. 1989. *Dryland: Environmental Management and Development.* Routledge, London.

Bedunah, D.J. and R.E. Sosebee. 1986. Influence of mesquite (*Prosopis gladulosa*) control on soil erosion on a depleted range site. *J. Soil Water Cons.* 41:131-135.

Blaikie, P.M. 1985. *The Political Economy of Soil Erosion in Developing Countries.* Longman, London.

Blaikie, P. 1989. Explanation and policy in land degradation and rehabilitation for developing countries. *Land Degrad. & Rehab.* 1:23-37.

Blaikie, P.M. and H. Brookfield. 1987. *Land Degradation and Society*, Methuen, New York.

Boahene, K. 1998. The challenge of deforestation in tropical Africa: reflections on its principal causes, consequences and solutions. *Land Degrad. & Develop.* 9:247-258.

Bravo, O., A. Quiroga, J. Silenza, S. Alcalde and E. Adema. 1995. Efectos del cultivos en franjas perennes sobre la estructura y carbono organico en suelos de la region semi-arid pampeana. *Revista de la Facultad de Agronomia la Plata* 71:15-20.

Brown, R.H. 1978. A difference in N use efficiency in C_3 and C_4 plants and its implications in adaptation and evolution. *Crop Sci.* 18:93-98.

Brown, R.H. 1994. The conservative nature of crop photosynthesis and the implications of carbon dioxide fixation pathways. In K.J. Boote, J.M. Bennett, T.R. Sinclair and G.M. Paulsen (eds), *Physiology and Determination of Crop Yield.* American Society of Agronomy, Madison, WI: 211-219.

Budyko, M.I. 1958. *The Heat Balance of the Earth's Surface.* Stepnova NA(trans) U.S. Dept. Comm., Washington, D.C.

Burgess, T.L. and A. Shmida. 1985. Succulent growth forms in arid environments. In E.E. Whitehead, C.F. Hutchinson, B.N. Timmermann and R.G. Varady (eds) *Arid Lands: Today and Tomorrow.* Westview Press: 363-395.

Carpenter, R.A. 1989. Do we know what we are talking about? *Land Degrad. & Rehab.* 1:1-3.

Chappell, A., W.N. Taylor and M. Charlton. 1998. Soil flux (loss and gain) in southwestern Niger, and its agricultural impact. *Land Degrad. & Develop.* 9:295-310.

Chaudhary, M.L., J.P. Singh and R.P. Narwal. 1981. Effect of long-term application of P, K and FYM on some soil chemical properties. *J. Indian Soc. Soil Sci.* 29:81-85.

Clark, J. and D. Noin (eds) 1997. *Population and Environment in Arid Regions.* MAB-UNESCO, vol. 19, The Parthenon Publishing Group, NY.

Conacher, A. 1998. Introduction. In A.J. Conacher and M. Sala (eds) *Land Degradation in Mediterranean Environments of the World.* J. Wiley & Sons, Chichester, U.K.: 171-173.

Darkoh, M.B.K. 1990. Trends in rural resource use and prospects for sustainable resource management in Kenya's arid and semi-arid lands. *Land Degrad. & Develop.* 2:177-190.

Darkoh, M.B.K. 1998. The nature, causes and consequences of desertification in the drylands of Africa. *Land Degrad. & Develop.* 9:1-20.

Dhir, R.P. and H.S. Mann. 1977. Soils. In *Desert Ecosystem and Its Improvement.* CAZRI Monograph 10, Jodhpur, India: 115-118.

Diaz, E., A. Roldan, V. Castillo and J. Albaladejo. 1997. Plant colonization and biomass production in a Xeric Torriorthent amended with solid refuse. *Land Degrad. & Develop.* 8:245-256.

Downtown, W.J.S. 1975. The occurrence of C_4 photosynthesis among plants. *Photosynthetica* 9: 96-105.

Dreccer, M.F. and R.S. Lavado. 1993. Influence of cattle trampling on preferential flow paths in alkaline soils. *Soil Use and Management* 9:143-148.

Dregne, H.E. 1976. *Soils of the Arid Regions.* Elsevier, Amsterdam.

Dregne, H.E. 1983. *Desertification of Arid Lands.* Harwood Academic Publishers, New York.

Dregne, H.E. 1995. Erosion and soil productivity in Australia and New Zealand. *Land Degrad. & Rehab.* 6:71-78.

Dregne, H.E. 1998. Desertification assessment. In R. Lal, W.H. Blum, C. Valentine and B.A. Stewart (eds) *Methods for Assessment of Soil Degradation*. CRC Press, Boca Raton, FL: 441-458.

Durrans, B. and R. Knox. 1982. *India: Past into Present*. The British Museum, London.

El-Baz, F. 1988. Origin and evolution of the desert. *Interdiscip. Sci. Rev.*, (Arrowsmith Ltd) 13(4):331-347.

Ellenberg, H. 1981. Reasons for stem succulents being present or absent in the arid region of the world. *Flora* 171:114-119.

Emerson, W.W. 1995. Water retention, organic C and soil texture. *Aust. J. Soil Res.* 33:241-351.

Essiet, E.U. 1990. A comparison of soil degradation under smallholder farming and large-scale irrigation land use in Kano State, Northern Nigeria. *Land Degrad. & Rehab.* 2:209-214.

FAO 1993. *Key Aspects of Strategies for the Sustainable Development of Drylands*. FAO, Rome.

FAO 1997a. *Dryland Development and Combating Desertification. Bibliographic Study of Experiences in China*. FAO Env. & Energy Paper 15, Rome.

FAO 1997b. *Irrigation Technology Transfer in Support of Food Security*. Water Reports 14, FAO, Rome.

FAO/UNESCO/WMO. 1977. World map of desertification. FAO, Rome.

Farrington, P. and R.B. Salama. 1996. Controlling dryland salinity by planting trees in the best hydrogeological setting. *Land Degrad. & Develop.* 7:183-204.

Galantini, J.A. and R.A. Rosell. 1997. Organic fractions, N,P, and S changes in Argentine semi-arid Haplustoll under different crop sequences. *Soil & Tillage Res.* 42:221-228.

Garg, V.K. 1998. Interaction of tree crops with a sodic soil environment: potential for rehabilitation of degraded environments. *Land Degrad. & Develop.* 9:81-93.

Gifford, R.M. 1974. A comparison of potential photosynthesis, productivity and yield of plant species with different photosynthetic metabolism. *Aust. J. Plant Phys.* 1:107-117.

Gifford, R.M., N.P. Cheney, J.C. Noble, J.S. Russell, A.B. Wellington and C. Zammit. 1992. Australian land use, primary production of vegetation and carbon pools in relation to atmospheric CO_2 levels. *Bureau of Rural Resources Proceedings* 14:151-187.

Glantz, M.H. and N.S. Orlovsky. 1986. Desertification: anatomy of a complex environmental process. In K.A. Dahlberg and J.W. Bennett (eds) *Natural Resources and People: Conceptual Issues in Inter-Disciplinary Research*. Westview Press, Boulder, CO: 213-229.

Gleick, P.H. 1985. Regional hydrologic impacts of global climate changes. In E.E. Whitehead, C.F. Hutchinson, B.N. Timmerman and R.G. Varady (eds) *Arid Lands: Today and Tomorrow*. Westview Press, Boulder, CO: 43-60.

Glenn, E., V. Squires, M. Olsen and R. Frye. 1993. Potential for C sequestration in the drylands., *Water Air and Soil Pollution* 70:341-355.

Grainger, A. 1990. *The Threatening Desert: Controlling Desertification*. Earthscan Publications, London.

Gupta, R.K. and I.P. Abrol. 1990. Salt-affected soils: their reclamation and management for crop production. In R. Lal and B.A. Stewart (eds) *Soil Degradation. Adv. Soil Sci.* 11:223-287.

Gupta, R.K., L.N. Harsh, K.A. Shankarnarayana, and B.D. Sharma. 1989. Wealth from wastelands. *Indian Farming* 38(11):18-19, 24.

Haque, I., J.M. Powell and S.K. Ehui. 1995. Improved crop-livestock production strategies for sustainable soil management in tropical Africa. In R. Lal and B.A. Stewart (eds) *Soil Management: Experimental Basis for Sustainability and Environmental Quality*. CRC/Lewis Publishers, Boca Raton, FL: 293-346.

Hare, F.K. 1985. *Climate Variations, Drought and Desertification*. WMO 653.

Harris, H. 1995. Long-term trials on soil and crop management at ICARDA. In R. Lal and B.A. Stewart (eds) *Soil Management: Experimental Basis for Sustainability and Environmental Quality*. CRC/Lewis Publishers, Boca Raton, FL: 447-472.

Hatch, M.D. 1987. C_4 photosynthesis: a unique blend of modified biochemisty, anatomy and ultrastructure. *Biochem. Biophys. Acta* 895:81-106.

Heathcote, R. 1983. *The Arid Lands: Their Use and Abuse*. Longman, London.

Hesla, B.I., L.I. Tieszen and S.K. Imbamba. 1982. A systematic survey of C_3 and C_4 photosynthesis in the Cyperaceae of Kenya, East Africa. *Photosynthetica* 16:196-205.

Higgit, D.L. and J.S. Rowan. 1996. Erosion assessment and administration in sub-tropical China: A case study from Fujian Province. *Land Degrad. & Develop.* 7:1-10.

Hillel, D. 1997. *Small-Scale Irrigation for Arid Zones: Principles and Options*. FAO Development Series 2, Rome.

Hinman, C.W. and J.W. Hinman. 1992. *The Plight and Promise of Arid Land Agriculture*. Columbia University Press, New York.

Houghton, J.T., J.G. Jenkins and J.J. Ephraums (eds) 1990. *Climate Change - the IPCC Scientific Assessment*. World Meteorological Organization (WMO), Cambridge Univ. Press, Cambridge.

IPCC 1996. Climate Change 1995. *Impacts, Adaptations and Mitigation of Climate Change: Scientific-Technical Analyses*. Working Group II. Cambridge Univ. Press, Cambridge, U.K.

Issar, A. and H. Tsoar. 1987. *Who Is to Blame for the Desertification of the Negev, Israel? The Influence of Climate Change and Climatic Variability on the Hydrologic Regime and Water Resources*. IAHS Publ. 168:577-583.

Jacobsen, T. and R.M. Adams. 1958. Salt and silt in ancient Mesopotamian agriculture. *Science* 128:1251-1258.

Jaiyeoba, I.A. 1998. Changes in soil properties related to conversion of woodland into pine and eucalyptus plantations, northern Nigeria. *Land Degrad. & Develop.* 9:207-215.

Jangir, R.P., B. Seeling and N.L. Joshi. 1997. Nutrient management reserach on pearl millet for the arid western plain zone of Rajsthan. In Proceedings of "From Research Station to Farmer's Field: Nutrient Management Research for Millet-Based Cropping Systems of Western Rajsthan." ICRISAT, Patancheru, India: 43-48.

Johnston, A.E. and J.K. Syers (eds) 1996. *Nutrient Management for Sustainable Crop Production in Asia*. CAB International, Wallingford, U.K.

Kolarkar, A.S., R.P. Singh and S. Singh. 1994. *Soils of the Indian Arid Zone. Sustainable Development of the Indian Arid Zone-A Research Perspective*. Scientific Publishers, Jodhpur, India: 25-31.

Krenzer, E.G., Jr., D.N. Moss and R.K. Crookston. 1975. Carbon dioxide compensation points of flowering plants. *Plant Physiol*. 56:194-206.

Ku, S-B. and G.E. Edwards. 1977. Oxygen inhibition of photosynthesis. II. Kinetic characteristics as affected by temperature. *Plant Physiol*. 59:991-999.

Kumar, M. and M.M. Bhandari. 1992. Impact of protection and free grazing on sand dune vegetation in the Rajsthan desert, India. *Land Degrad. & Rehab*. 3:215-228.

Kumar, M. and M.M. Bhandari. 1993. Human use of sand dune ecosystem in the semi-arid zone of the Rajsthan desert, India. *Land Degrad. & Rehab*. 4:21-36.

Kumar, A.B., K.N. Tiwari, R.S. Dwivedi and D. Karunakar. 1997. Spectral behaviour and spectral separability of eroded lands using multi-sensor data. *Land Degrad. & Develop*. 8:27-38.

Lal, R. 1990. *Soil Erosion in the Tropics: Principles and Management*. McGraw Hill, New York.

Lal, R. 1994a. Soil erosion by wind and water: problems and prospects. In R. Lal (ed) *Soil Erosion Research Methods*. St. Lucie Press, Delray Beach, FL: 1-10.

Lal, R. 1994b. Global overview of soil erosion. In Global Overview of Soil Erosion. *Soil Sci. Soc. Amer*. Special Publication No. 41, Madison, WI: 39-51.

Lal, R. 1995. Global soil erosion by water and C dynamics. In R. Lal, J. Kimble, E. Levine and B.A. Stewart (eds) *Soils and Global Change*. CRC/Lewis Publishers, Boca Raton, FL: 131-141.

Lal, R., G.F. Hall and F.P. Miller. 1989. Soil degradation. I. Basic Processes. *Land Degradation & Rehabilitation* 1:51-69.

Lal, R., J. Kimble, R.F. Follet and C.V. Cole. 1998. *The Potential of U.S. Cropland to Sequester Carbon and Mitigate the Greenhouse Effect*. Ann Arbor Press, Chelsea, MI.

Lamachère, J.M. and G. Serpantie. 1991. Valorisation agricole des eaux de ruissellement et lutte contre l'érosion sur champs de mil en zone soudano-sahélienne. Région de Bidi, Yatenga, Burkina Faso. *Bull. Réseau Erosion* 11:88-104. (ORSTOM, Montpellier).

Laryea, K.B., M.M. Andres and P. Pathak. 1995. Long-term experiments on Alfisols and Vertisols in the semi-arid tropics. In R. Lal and B.A. Stewart (eds) *Soil Management: Experimental Basis for Sustainability and Environmental Quality*. CRC/Lewis Publishers, Boca Raton, FL: 267-292.

Le Houérou, H.N. 1975. *Science power and desertization*. Meeting on desertization. Dept. Geogr., Univ. Cambridge, 22-28 Sept. 1975: 26pp.

Lettau, H. 1969. Evapotranspiration climatonomy 1. A new approach to numerical prediction of monthly evapotranspiration, runoff and soil moisture storage. *Mon. Weather Rev.* 97:691-699.

Lewis, L.A. and V. Nyamulinda. 1996. The critical role of human activities in land degradation in Rwanda. *Land Degrad. & Develop.* 7:47-56.

Leys, J. and G. McTainsh. 1994. Soil loss and nutrient decline by wind erosion-cause for concern. *Aust. J. Soil and Water Conserv.* 7:30-35.

Li, Zhong Pei, Li Li Cheng, Lin Xin Xiong, Z.P. Li , L.L. Cheng and X.X. Lin. 1994. Studies on the characterisitcs of organic matter in degraded red soil and the effect of application of organic manure on it. *Soils* 26:70-76.

Lomte, M.H., M. Ateeque, P.R. Bharambe and P.K. Kawarkhe. 1993. Influence of sorghum-legume association on physico-chemical properties of soil. *J. Maharashtra Agric. Univ.* 18:388-390.

Mabbutt, J.A. 1984. A new global assessment of the status and trends of desertification. *Env. Cons.* 11:100-113.

Mainguet, M. 1991. *Desertification: Natural Background and Human Mismanagement*. Springer-Verlag, Berlin.

Mando, A. 1997. Effect of termites and mulch on the physical rehabilitation of structurally crusted soils in the Sahel. *Land Degrad. & Develop.* 8:269-278.

Mendoza, M. 1990. Global assessment of desertification: world atlas of thematic indicators of desertification: proposal document. In *Desertification Revisited: Proceedings of an Adhoc Consultative Meeting on the Assessment of Desertification*. UNEP DC/PAC, Nairobi: 289-294.

Michels, K., M.V.K. Sivakumar and B.E. Allison. 1995. Wind erosion control using crop residue. I. Effects on soil flux and soil properties. *Field Crops Res.* 40:101-110.

Middleton, N.J. and D.S.G. Thomas. 1992. *World Atlas of Desertification*. Edward Arnold/UNEP, Sevenoaks.

Miglierina, A.M., M.R. Landriscini, J.A. Galantini, J.O. Iglesias, A. Glave and J.F. Gallardo-Lancho. 1993. Fifteen years of crop rotation and fertilization in the semi-arid pampas region. I. Effect on the chemical properties of the soil. El estudio del suelo y de su degradacion en relacion con la desertification. Actas del 12 Congreso Latinamericano de la Ciencia del Suelo, Sevilla (Espana), 19-26 Sept., 1993: 751-758.

Miglierina, A.M., J.A. Galantini, J.O. Iglesias, R.A. Rosell and A. Glave. 1996. Crop rotation and fertilization in production systems of the semi-arid region of Buenos Aires. *Revista de la Facultad de Agronomia, Universidad de Buenos Aires* 15:9-14.

Mishra, M.M., S. Neelakantan and K.C. Khandelwal. 1974. Buildup of soil organic matter continuous manuring and cropping practices. *J. Res. Hayana Agric. Univ.* 4:224-230.

Mohammed, S.O., A. Farshad and J. Farifteh. 1996. Evaluating land degradation for assessment of land vulnerability to desert conditions in the Sokoto area, Nigeria. *Land Degrad. & Develop*. 7:205-216.

Monger, H.C. and R.A. Gallegos. 1999. Biotic and abiotic processes and rates of pedogenic carbonate accumulation in the southwestern United States - relationship to atmospheric CO_2 sequestration. In R. Lal, J. Kimble, H. Eswaran and B.A. Stewart (eds) *Pedogenic Carbonates and C Sequestration*. Ann Arbor Press, Chelsea, MI (in press).

More, S.D. 1994. Effect of farm wastes and organic manures on soil properties, nutrient availability and yield of rice-wheat grown on sodic Vertisol. *J. Indian Soc. Soil Sci*. 42:253-256.

Muthuvel, P., R.K. Pallikondaperumal, R. Sivasamy, V. Subramanian, V.S. Sreeramulu. 1989. Soil fertility under continuous rotational cropping of cotton-pearl millet in a dryland Vertisol. *Madras Agric*. J. 76:189-191.

Nambiar, K.K.M. 1995. Major cropping systems in India. In V. Barnett, R. Payne and R. Steiner (eds) *Agricultural Sustainability, Economic, Environmental and Statistical Considerations*. J. Wiley & Sons, Chichester, U.K.: 133-170.

Niemeijer, D. 1998. Soil nutrient harvesting in indigenous *Teras* water harvesting in eastern Sudan. *Land Degrad. & Develop*. 9:323-330.

Noin, D. and J.I. Clarke. 1997. Population and environment in arid regions of the world. In J. Clarke and D. Noin (eds) *Population and Environment in Arid Regions*. MAB/UNESCO, vol. 19, The Parthenon Publishing Group, New York: 1-18.

Nye, P.H. and D.J. Greenland. 1962. *Soils under Shifting Cultivation*. Commonwealth Agric. Bureau, Harpendon, England.

Oba, G. 1992. Effects of controlled grazing on a degraded swarf shrub, annual grass semi-desert vegetation type of northwestern Kenya. *Land Degrad. & Rehab*. 3:199-214.

Odingo, R.S. 1990. The definition of desertification and its programmatic consequences for UNEP and the international community. *Desertification Control Bulletin* 18:31-70.

Oldeman, L.R. 1994. The global extent of soil degradation. In D.J. Greenland and I. Szabolcs (eds) *Soil Resilience and Sustainable Land Use*. CAB International, Wallingford, U.K.: 99-117.

Oldeman, L.R. and G.W.J. Van Lynden. 1998. Revisiting the GLASOD methodology. In R. Lal, W.H. Blum, C. Valentine and B.A. Stewart (eds) *Methods for Assessment of Soil Degradation*. CRC Press, Boca Raton, FL: 423-440.

Omar, S.A.S., T. Madouh, I. El-Bagouri, Z. Al-Mussalem and H. Al-Telaihi. 1998. Land degradation factors in arid irrigated areas: the case of Wafra in Kuwait. *Land Degrad. & Develop*. 9:283-294.

Patil, S.G., M. Hebbara and S.B. Devarnavadagi. 1996. Screening of multipurpose trees for saline Vertisols and their bioameliorative effects. *Annals Arid Zone* 35:57-60.

Pearcy, R.W. 1983. The light environment and growth of C_3 and C_4 tree species in the understory of a Hawaiian forest. *Oecologia* 58:19-25.

Pearcy, R.W. and H.W. Calkin. 1983. Carbon dioxide exchange of C_3 and C_4 tree species in the understory of a Hawaiian forest. *Oecologia* 58:26-32.

Perkins, J.S. and D.S.G. Thomas. 1993. Spreading deserts or spatially confined environmental impacts? Land degradation and cattle ranching in the Kalahari desert of Botswana. *Land Degrad. & Rehab.* 4:179-194.

Pieri, C. 1991. *Fertility of Soils: A Future for Farming in the West African Savannah.* Springer Verlag, Berlin, 348pp.

Pieri, C. 1995. Long-term soil management experiments in semi-arid Francophone Africa. In R. Lal and B.A. Stewart (eds) *Soil Management: Experimental Basis for Sustainability and Environmental Quality*. CRC/Lewis Publishers, Boca Raton, FL: 225-266.

Pieri, C., J. Dumanski, A. Hamblin and A. Young. 1995. *Land Quality Indicators.* World Bank Discussion Paper 315. World Bank, Washington, DC.

Qadir, M., R.H. Qureshi, N. Ahmad and M. Ilyas. 1996. Salt-tolerant forage cultivation on a saline-sodic field for biomass production and soil reclamation. *Land Degrad. & Develop.* 7:11-18.

Ragab, M., H. Lieth and A.A. Al-Masoom. 1990. Distribution pattern of soil microbial population in salt-affected soils. In *Towards the Rational Use of High-Salinity Tolerant Plants. Vol. 1. Deliberations About High Salinity Tolerant Plants and Ecosystems.* Proc. of the First ASWAS Conference, 8-15 December 1990, Al Ain, UAE 1993: 467-472.

Raji, B.A., I.E. Esu, V.O. Chude, J.J. Owonubi and T. Kparmwang. 1996. Properties, classification, and management implications of soils of Illela sand dunes. *J. Sci. Food & Agric.* 71:425-432.

Rao, B.R.M., R.S. Dwivedi, K. Sreenivas, Q.I. Khan, K.V. Ramana, S.S. Thammappa and M.A. Fyzee.1998. An inventory of salt-affected soils and waterlogged areas in Nagarjunsagar Canal Command Area of southern India using space-borne multi-spectral data. Land Degrad. & Develop. 9: 357-368.

Ravindranath, N.H. and B.S. Somashekhar. 1995. Potential and economics of forestry options for carbon sequestration in India. *Biomass and Bioenergy* 8:323-336.

Roose, E.J. 1996. *Land Husbandry: Components and Strategy.* FAO Soils Bulletin 70, Rome, Italy, 380pp.

Ruecker, G., P. Sachad, M.M. Alcubilla and C. Ferrer. 1998. Natural regeneration of degraded soils and site changes on abandoned agricultural terraces in Mediterranean Spain. *Land Degrad. & Develop.* 9:179-188.

Ruhal, D.S. and M. Singh. 1982. Effect of long-term application of farmyard manure and nitrogen. I. Crop yields and accumulation of organic carbon. *J. Res. Haryana Agric. Univ.* 12:272-281.

Sage, R.F. and R.W. Pearcy. 1987. The nitrogen use efficiency of C_3 and C_4 plants. II. Leaf nitrogen effects on the gas exchange characteristics of *Chenopodium album* and *Amaranthus netroflexus*. *Plant Physiol.* 84:959-963.

Salisbury, F.B. and C.W. Ross. 1978. *Plant Physiology*, 2nd Edition. Wadsworth Publishing Co., Belmont, CA: 136-173.

Salisbury, F.B. and C.W. Ross. 1992. *Plant Physiology*, 4th Edition. Wadsworth Publishing Co., Belmont, CA: 255-265.

Sanchez, P.A. and R.J. Buresh (eds) 1997. Replenishing soil fertility in Africa. *Soil Sci. Soc. Am.* Special Publication No. 51, Madison, WI.

Schlesinger, W.H. 1990. Evidence from chronosequence studies for a low carbon-storage potential of soils. *Nature* 348 (6298): 232-234.

Schlesinger, W.H. 1997. *Biogeochemistry: An Analysis of Global Change*. Academic Press, New York.

Seitlheko, E.M., B.L. Allen and D.B. Wester. 1993. Effect of three grazing intensities on selected soil properties in semi-arid west Texas. *African J. Range & Forage Sci.* 10:82-85.

Shahin, R.R., M.A. El-Meleigi, A.A. Al-Rokiba and A.M. Eisa. 1998. Changes in organic matter and fertility of sandy soils as affected by wheat crop rotation. *Bulletin Fac. Agric. Univ. Cairo* 49:169-184.

Singh, B. 1989. Rehabilitation of alkaline wasteland on the Gangetic alluvial plains of Uttar Pradesh, India, through afforestation. *Land Degradation and Rehabilitation* 1:305-310.

Singh, G., I.P. Abrol and S.S. Cheema. 1988. Agroforestry on alkali soil: Effect of planting methods and amendments on initial growth, biomass accumulation and chemical composition of mesquite (*Prosopis juliflora*) with inter-space planted with and without karnal grass (*Diplachne fusca*). *Agroforestry Systems* 7:135-160.

Singh, G, I.P. Abrol and S.S. Cheema. 1989. Effects of gypsum application on mesquite (*Prosopis juliflora*) and soil properties in an abandoned sodic soil. *Forest Ecology and Management* 29:1-14.

Singh, G., H.S. Gil and G. Singh. 1992. Ameliorative effect of tree species on characteristics of sodic soils at Karnal. *Indian J. Agric. Sci.* 62:144-146.

Singh, G., and H.C. Goma. 1995. Long-term soil fertility management experiments in East Africa. In R. Lal and B.A. Stewart (eds), *Soil Management: Experimental Basis for Sustainability and Environmental Quality*. CRC/Lewis Publishers, Boca Raton, GL: 347-384.

Singh, G., N.T. Singh and I.P. Abrol. 1994. Agroforestry techniques for the rehabilitation of degraded salt-affected lands in India. *Land Degrad. & Rehab.* 5:223-242.

Singh, B.R. and H.C. Goma. 1995. Long-term soil fertility management experiments in East Africa. In R. Lal and B.A. Stewart (eds) *Soil Management: Experimental Basis for Sustainability and Environmental Quality*. CRC/Lewis Publishers, Boca Raton, FL: 347-384.

Singh, G. and G. Singh. 1995. Long-term effects of amendments on *Prosopis juliflora* and soil properties of a highly alkali soil. *J. Tropical Forest Sci.* 8:225-239.

Singh, G. and G. Singh. 1996. Effect of site preparation technique on *Prosopis juliflora* in an alkali soil. *Forest Ecology & Management* 80:267-278.

Singh, G., J.C. Dagar and N.T. Singh. 1997. Growing fruit trees in highly alkali soils - a case study. *Land Degrad. & Develop.* 8:257-268.

Smyth, A.J. and J. Dumanski. 1993. *FESLM: An International Framework for Evaluating Sustainable Land Management*. World Soil Resources Report 73. Food and Agriculture Organization, Rome.

Squires, V., E.P. Glenn and A.T. Ayoub. 1995. Combating global climate change by combating land degradation. *Proceedings* of a workshop held in Nairobi, Kenya, 4-8 September, 1995. UNEP, Nairobi, Kenya.

SSSA 1995. "SSSA Statement on Soil Quality." *Agronomy News* (June 1995). Soil Science Society of America, Madison, WI.

Sterk, G., L. Herrmann and A. Bationo. 1996. Wind-blown nutrient transport and soil productivity changes in southwest Niger. *Land Degrad. & Develop*. 7:325-336.

Stiles, D. 1988. Arid land plants for economic development and desertification control. *Desertification Control Bull*., UNEP, Nairobi, Kenya 17:18-21.

Sumner, M.E. and R. Naidu (eds). 1998. *Sodic soils: distribution, properties, management and environmental consequences*. Oxford University Press, New York.

Sumner, M.E., P. Rengasamy and R. Naidu. 1998. Sodic soils: a reappraisal. In M.E. Sumner and R. Naidu (eds) *Sodic soils: distribution, properties, management and environmental consequences*. Oxford University Press, New York: 1-17.

Swift, M.J., P.D. Seward, P.G.H. Frost, J.N. Qureshi and F.N. Muchena. 1994. Long-term experiments in Africa: developing a database for sustainable landuse under global change. In R.A. Leigh and A.E. Johnston (eds) *Long-term Experiments in Agricultural and Ecological Sciences*. CAB International, Wallingford, UK: 229-251.

Taiz, L. and E. Zeiger. 1998. *Plant Physiology,* 2nd Edition. Sinauer Associates, Sunderland, MA: 195-228.

Thomas, D.S.G. and N.J. Middleton. 1994. *Desertification: Exploding the Myth*. J. Wiley & Sons, Chichester, UK.

Traoré, S. and P.J. Harris. 1995. Long-term fertilizer and crop residue effects on soil and crop yields in the Savanna region of Côte d' Ivoire. In R. Lal and B.A. Stewart (eds) *Soil Management: Experimental Basis for Sustainability and Environmental Quality*. CRC/Lewis Publishers, Boca Raton, FL: 141-180.

UNCED 1992. *Earth Summit Agenda 21: Programme of action for sustainable development*. UNEP, New York.

UNEP 1977. World map of desertification. UN Conference on desertification. 29 August - 9 September 1977. Document A/Conf. 74/2, Nairobi, Kenya.

UNEP 1984a. *General Assessment of the Progress in the Implementation of the Plan of Action to Combat Desertification 1977-84*. NIP, Nairobi, Kenya.

UNEP 1984b. *Activities of the United Nations Environment Programme in the Combat against Desertification*. Report prepared by the Desertification Control/Programme Activity Center, UNEP, Nairobi, Kenya.

UNEP 1987. *Rolling Back the Desert*. UNEP/DC/PAC, Nairobi, Kenya.

UNEP. 1990. *Desertification Revisited*, UNEP/DC/PAC, Nairobi, Kenya.

UNEP 1991. *Status of Desertification and Implementation of the United Nations Plan of Action to Combat Desertification*. UNEP, Nairobi, Kenya.

UNEP 1992. *World Atlas of Desertification*. Edward Arnold, Sevenoaks. UNEP, Nairobi, Kenya.

UNESCO 1983. *World Distribution of Arid Regions*. UNESCO, New York.

Van Dijk, J.A. 1997. Indigenous soil and water conservation by Teras in eastern Sudan. *Land Degrad. & Develop*. 8:17-26.

Verstraete, M.M. 1986. Defining desertification: a review. *Climatic Change* 9:5-18.

Wakeel, A. El and A. Astartke.1996. Intensification of agriculture on Vertisols to minimize land degradation in parts of the Ethiopian Highlands. *Land Degrad. & Develop*. 7:57-68.

Waller, S.S. and J.K. Lewis. 1979. Occurrence of C_3 and C_4 photosynthetic pathways in North American grasses. *J. Range Management* 32:12-29.

Whitehead, E.E., C.F. Hutchinson, B.N. Timmermann and R.G. Varday (eds) 1985. *Arid Lands: Today and Tomorrow*. Westview Press, Boulder, CO.

Wiggs, G.F.S., I. Livingstone, D.S.G. Thomas and J.E. Bullard. 1994. Effect of vegetation removal on airflow patterns and dune dynamics in the southwest Kalahari desert. *Land Degrad. & Rehab*. 5:13-24.

Wilson, J.R. and R.H. Brown. 1983. Nitrogen response of *Panicum* species differing in CO_2 fixation pathways. I. Growth analysis and carbohydrate accumulation. *Crop Sci*. 23:1148-1153.

Winter, K. and J. Troughton. 1978. Photosynthetic pathways in plants of coastal and inland habitats of Israel and the Sinai. *Flora Bot*. 167:1-34.

World Bank. 1997. *Rural Development: From Vision to Action*. Environmentally and Socially Sustainable Development Studies and Monographs Series 12. World Bank, Washington, DC.

Zaimeche, S.E. and K. Sutton. 1990. The degradation of the Algerian environment through social development in the 1980s. *Land Degrad. & Develop*. 2:317-324.

Zhao, S. 1986. *Physical Geography of China*. Scientific Press, Beijing, China, Wiley, New York.

APPENDIX I: Land Quality Indicators (LQIs)

Although a single indicator of land quality is not realistic, a very large number reflecting all possibilities is also not useful. To help resolve this problem, a core set of strategic land quality indicators is described below.

(a) Usable core LQIs:

(i) *Nutrient balance*. It involves nutrient stocks and flows as related to different land management systems used by farmers in specific agroecological zones (AEZs) and specific countries.

(ii) *Yield gap*. It involves current yields, yield trends, and actual:potential farm-level yields in cereal equivalents.

(iii) *Land use intensity*. This index is based on the impacts of agricultural intensification on land quality. Intensifcation may involve increased cropping, more value-added production, and increased amounts and frequency of inputs, i.e., management practices adopted by farmers in the transition to intensification.

(iv) *Land use diversity (agrodiversity)*. It involves the extent of diversification of production systems over the landscape, including livestock and agroforestry system and reflects the degree of flexibility (and resilience) of regional farming system and their capacity to absorb shocks and respond to opportunities.

(v) *Land cover*. The indicator includes the extent, duration, and timing of vegetative cover in the land during major erosive periods of the year. Land cover is a surrogate for erosion and, along with land use intensity and diversity, offers increased understanding on issues of desertification.

(b) Core LQIs being researched:

(i) *Soil quality*. It involves the conditions that make the soil a living body, i.e., soil health. The indicators will be based on soil organic matter, particularly the dynamic (microbiological) carbon pool most affected by environmental conditions and land use change.

(ii) *Land degradation* (erosion, salinization, compaction, organic matter loss). These processes have been much researched and have a strong scientific base, but reliable data on extent and impacts are often lacking.

(iii) *Agrobiodiversity*. This concept involves managing the gene pools utilized in crop and animal production, but also soil micro and meso biodiversity important for soil health. On a macro scale, it involves integrated landscape management including maintenance of natural habitiat, as well as managing the coexistence of wildlife in agricultural areas.

(c) Core LQIs being developed by other sectors:

(i) Water quality

(ii) Forest land quality

(iii) Range land quality

(iv) Land contamination/pollution.

These strategic indicators are the biophysical components of sustainable land management. Although useful in their own right, they must still be complemented with indicators of the other pillars of sustainable land management: economic viability, system resilience, and social equity and acceptability.

APPENDIX II: The Integration of Policies and National Development Programs to Better Manage Global Carbon Stocks and Flows

The opportunities for enhanced carbon sequestration must be viewed in the context of resolving local, regional, and national environmental problems, as well as global environmental issues such as climate change, loss of biological diversity, land degradation and desertification, deforestation and unsustainable forestry, and degradation of fresh and marine waters. The primary focus is on local and national development priorities, but it is clear that national development will not be sustainable unless degradation of the global commons is also addressed. This is especially important for developing countries, because of their comparatively lower capacity to mitigate the impacts of global change.

International assistance efforts are available to help countries adjust to the opportunities related to management of global environment change, when such action benefits or is at no additional cost to development. Realizing these opportunities requires action on:

- reducing negative impacts of global environment change on development and developing measures that can be taken to mitigate these impacts
- creating opportunities to take economic advantage of emerging markets for global environment goods and services
- developing procedures to comply with the global environment conventions, to the extent that such actions are "win-win" or fully compensated for any costs that exceed domestic benefits.

Global Environment Conventions Related to Management of Carbon Stocks and Flows

Several of the international conventions impact directly on management of global carbon stocks and flows, whereas others have indirect impacts. The conventions that impact *directly* are:

(i) *United Nations Framework Convention on Climate Change (UNFCCC) and the Kyoto Protocol:* This convention emphasizes reduction of atmospheric greenhouse gas emissions. In terms of management of carbon sequestration, the current focus is on reduction of deforestation and reforestation efforts, but future activities may include opportunities to sequester carbon in the soil through improved land management. The goal of the UNFCCC is "the stabilization of greenhouse gas concentrations in the atmosphere at a level that would prevent dangerous anthropogenic interference with the climate system." The UNFCCC provides a framework under which countries can begin to address the threat of climate change, but does not contain binding obligations to limit emissions of greenhouse gases. At the Third Conference of the Parties to the UNFCCC in Kyoto, Japan, in December 1997, the Parties adopted the "Kyoto Protocol." The Protocol sets binding emissions reduction for industrialized countries and those with economies in transition. Overall emissions of greenhouse gases have to be reduced by at least 5% below 1990 levels in the commitment period 2008 to 2012. Developing countries do not assume any emission reduction obligations at this time, although they have responsibilities such as formulating regional emission reduction programs that take account of the socioeconomic conditions of each country, and implementing other measures, including market-based mechanisms, to address climate change.

(ii) *United Nations Convention to Combat Desertification (UNCCD)*: This convention links directly with land degradation, and thus provides opportunities for carbon sequestration through improved land management, involving both reduced oxidation of soil carbon as well as increased storage of carbon in below- and above-ground biomass. The UNCCD obligates parties to (a) give due priority to combating desertification and mitigating the effects of drought; (b) establish strategies, priorities, and policies to combat desertification and mitigate the effects of drought; (c) address the underlying causes of desertification and pay special attention to the socioeconomic factors contributing to desertification processes; (d) promote awareness and facilitate participation of local populations in efforts to combat desertification and mitigate the effects of drought; (e) provide an enabling environment by strengthening relevant existing legislation and enacting new laws and establishing long-term policies and action programs. Over 130 countries have ratified the UNCCD. Implementation is carried out though collaboration among international organizations, governments, the private sector, the civil society, affected populations and NGOs.

(iii) *Convention on Biological Diversity (CBD)*: The CBD obligates Parties to "conserve and sustainably use their own biological diversity in accordance with their particular conditions and capabilities." In addition, it calls "for the equitable sharing of benefits from the utilization of genetic resources." However, mechanisms to promote these goals have only been developed to a limited extent, and the CBD does not translate these general objectives into more specific targets around which resources could be mobilized and progress assessed. The CBD requires each Party to develop "as far as possible and appropriate" national strategies, plans or programs for the conservation and sustainable use of biological diversity and to integrate these plans and programs into relevant sectoral and cross-sectoral planning and practice. At the last meeting of the Conference of Parties of the CBD, it was decided to give special attention to land degradation for this biennium.

The following convention *indirectly impacts* the management of global carbon stocks and flows:

(iv) *International Legal Framework on Water Resources*: Currently, the only legally binding global agreement on management of water resources is the U2V Convention on the Law of the Sea, which is primarily economic in nature. At the regional level, there are a plethora of agreements for the management of international river basins, transboundary lakes, and enclosed and marginal seas, including the ten regional sea conventions for the protection of the marine environment under UNEP's Regional Seas Programme. An international consensus has emerged on the principles of sound fresh and marine water resources management, based on several major international conferences and conventions. The Convention on the Law of Non-Navigational Uses of International Water Courses was adopted by the UN General Assembly in 1997. Earlier conventions include the International Conference on Water and Environment in Dublin, the Earth Summit and Agenda 21 in 1992, the Convention on Biological Diversity, with its Jakarta Mandate on Coastal and Marine Biodiversity, and most recently, the Global Programme of Action (GPA) for the Protection of the Marine Environment Against Land Based Activities, adopted in 1995. The common themes are the need for integrated water resources planning and management in river basins and the coastal zone, linkages between land based activities on water quality and supply, and the cascading effects of impacts in the upper watershed on productive sectors in the coastal zone and marine environment.

Opportunities, Tradeoffs and Constraints to Achieving Improved Management of Global Carbon Stocks and Flows

Global environment problems do not stop at national, geo-political, or economic boundaries, and there is increased understanding that degradation of the "global commons" will cause collective suffering in both developed and developing countries if current trends of global environment deterioration continue. However, there are difficult decisions in defining the appropriate level of effort that should be devoted to global environmental management. In particular, developing countries perceive real and critical tradeoffs among:

- meeting pressing short term needs such as food, water supply, and sanitation services for the poor, and investing in environmental management for the medium and long term
- harvesting natural resources to generate foreign exchange urgently required to purchase essential imported supplies, or simply to meet the basic needs of the urban and rural poor, and the need to manage these resources for long term, sustainable use
- within the broader environmental agenda, expenditures on local and regional pollution abatement and expenditures to take actions on global environmental change and its local impacts.

Getting the appropriate balance is not easy, especially when political regimes tend to have a short term time horizon, but at the same time, these questions reach to the core of defining sustainable development in practical terms. In many cases, the Global Environmental Facility (GEF) is an important partner in resolving many of these and related issues.

These perceived tradeoffs impact the importance that countries attach to environmental priorities generally, and global ones in particular. In addition, developing countries face a number of specific constraints that limit their ability to take action on these priorities:

- Limited economic incentives in individual countries to take action on the global environment, in the absence of markets for trading certain global environmental goods and services.
- Weak policy frameworks in sectors, such as energy, forestry, water resources, and environmental management, combined with insufficient political commitment; this limits the incentives and scope for effective action on environmental issues generally.
- Institutional weaknesses, particularly the lack of technical skills on environmental issues and management of environmental issues; as a result, global environmental externalities are rarely integrated into sector development strategies.

Recent Initiatives in the World Bank to Strengthen Sector Policies in Relation to Global Environmental Issues

Key sector polices and strategies for energy, forestry and rural development are being revised to make them fully responsive to environmental issues, including global environmental issues.

- The Bank's *energy policies* support lending only when countries commit to improving energy efficiency through policy reforms and promote renewable energy. While good progress has been made on the sector reform agenda, the volume of lending for energy efficiency, renewables and rural energy remains modest. This is largely because environmentally friendly energy projects are more labor-intensive, borrower commitment to fundamental reform is often weak, and energy efficiency and renewable energy are relatively new business products for which skills and experience are scarce.

- The Bank's *Forestry Policy* (1991) precludes financing of commercial logging operations in moist tropical forests, and increasingly recognizes the roles of the private sector and local people in forestry and conservation management. Also, lending operations in the forest sector are conditional on government commitment to sustainable management and conservation-oriented forestry. A review of the implementation of the forest policy is underway, with the intention of developing a forest sector strategy with monitorable progress indicators.
- The *Bank's Rural Development Strategy* (1997) recognizes the link between rural development and global environment issues. To help build this link into Bank operations, global environmental concerns are being integrated into sector activities through a Sustainable Land and Crop Management thematic team. Agriculture sector performance indicators are being revised to include global environmental concerns, and the issues of land degradation and desertification are given special importance. With IFAD and UNDP, the Bank Group is helping to establish mechanisms for mobilizing financial resources for implementation of the Desertification Convention.
- The Bank's *Water Resources Policy* (1993) acknowledges that protection of water resources is essential for sustainable development and calls for a comprehensive management approach. It emphasizes rigorous attention to maintaining biodiversity and protecting ecosystems in water project design and implementation.

New Opportunities, Products and Mechanisms for Improved Management of Global Carbon Stocks and Flows

New initiatives and mechanisms provide opportunities for improved management of global carbon.

- The establishment of a Prototype Carbon Fund (PCF) is a new initiative being examined to link private sector funding with international assistance to develop a market mechanism for the transfer of carbon offsets. The PCF would fund projects by covering the cost of emission reduction measures above a baseline, through contributions made by industrial countries and private sector companies. Some investments by the private sector have already been made under the umbrella of this activity, but the mechanisms to manage the fund are yet to be developed.
- Some strategic partnerships for renewable energy technologies are under development. An example is the evolving Bank-GEF partnership to increase the competitiveness of new and renewable technologies that do not emit greenhouse gases.
- The Bank has recently launched two Market Transformation Initiatives with coalitions of industries and NGOs that aim to promote market-led change in the adoption of ecologically sustainable forest and marine production and harvesting practices.
- Several strategic partnerships have been established to support conservation and sustainable use of biodiversity in forestry. Among them are (a) The World Bank-World Wide Fund for Nature Alliance for forest conservation and sustainable forest management; (b) Forest Trends, an independent non-profit NGO (together with World Resources Institute), to provide information on forest management and serve as a forum for resolving contentious issues; (c) a study on four critical terrestrial ecosystems (with Conservation International) as a first step in the detailed design of the Critical Ecosystems Partnership Fund; and (d) an evolving World Bank-International

Federation of Agricultural Producers (IFAP) partnership to promote sustainable agricultural development and sustainable land management.

The above are some of the major initiatives intended to help developing countries overcome constraints in integrating the global environment in national sustainable development. They are mostly at early stages of establishing dialogue and identifying opportunities, and their impact on development investments in energy, forestry, rural development will have to be judged later.

Challenges Ahead

Notwithstanding the progress that has been made, a number of major challenges remain. It is not clear what priority developing countries will assign to investments in "win-win" environmental opportunities, i.e., those which benefit the global environment and the domestic economy simultaneously. These opportunities must be vetted against the many competing claims that developing countries face. There is continual tension between the need for high rates of return on scarce development capital and the frequently lower rates of return on longer term investments required to promote sustainable use of key natural capital assets such as soil, water, forests, and biodiversity. Hence, even when the "win-win" investments meet investment objectives, they may still be discriminated against because of the longer term investments required for such programs. Some forms of preferential lending terms, and other financial instruments such as taxation, set-asides, etc. may be necessary to buy down the transaction costs to introducing climate friendly renewable energy technologies, soil carbon sequestration programs, and other related activities.

Other major challenges are to strengthen national and international capacity:

- To identify "win-win" opportunities in sectoral reform and investment programs, for capturing global benefits under existing funding mechanisms
- To build technical capacity and supporting networks of professionals within and between countries to pursue and analyze these opportunities
- To strengthen the administrative and managerial functions in key agencies to manage these initiatives.

The objective is not to promote centralized "global environmental management capacity" within environmental ministries, but rather to help senior officials, planners and technicians in agricultural, forestry, and the energy/transportation sectors to recognize the importance of conservation and improved management of global carbon stocks and flows, and to see the opportunities to use environment-friendly policies to complement their traditional goals.

Finally, and in accord with current major efforts to decentralize decision making to regional and local levels, a major challenge is to identify the opportunities and develop the appropriate procedures for community-based, "bottom-up" rather than "top-down" delivery of programs with global environmental benefits. In the final analyses, improved management of the global commons will be achieved through collective actions of decision makers, such as farmers and forest managers who make daily decisions in their local areas. Procedures must be developed to empower these individuals and groups to make the appropriate choices, but also to assume responsibility for their decisions. Although the impacts of individual actions may appear to be inconsequential, these actions are repeated over and over on the landscape, and collectively they can impact global carbon stocks and flows.

COMMENTARIES ON "DESERTIFICATION CONTROL TO SEQUESTER C AND MITIGATE THE GREENHOUSE EFFECT"

The commentaries that follow have been edited slightly so as to highlight their key points. Although some or most of the commentators' suggestions were adopted by the white paper authors in their revisions, we have left them intact so the reader can gain an understanding of the discussions and dynamics of the process used at the workshop.

PIERRE CROSSON, *Resources for the Future*

Introduction

This is an interesting paper done at a high professional level, the kind of thing one would expect from Rattan Lal and his collaborators here. Yet the paper suffers from some significant weaknesses. It is very ambitious, excessively so in my judgment. The sections on Indicators of Land Quality to Monitor Trends in Carbon Sequestration and The Integration of Policies and National Development Programs to Better Manage Carbon Stocks and Flows raise issues of theory, data and policy of enormous complexity which no one, not even authors as capable as these, could possibly handle adequately in a single paper. Moreover, these sections detract from the main topic of the paper, namely, what might be the contribution of desertification control in dry areas to reduction in carbon emissions from the soil? My comments here are focused on the treatment of this topic.

Potential Reductions in Soil Carbon Emissions

The authors identify six sources of potential importance in reducing soil carbon emissions in dry areas. Four of these—erosion control, restoration of eroded land, reclamation of salt-affected soils, and a carbon credit for biofuel production—account for 99% of the estimated potential reduction in C emissions: reclamation of salt-affected soils for 27%, erosion control for 25%, restoration of eroded land for 24%, and the biofuel credit for 23%. My comments focus on these four options.

The estimate of potential reductions in soil carbon emissions from these sources is significant—about 1 billion metric tons per year. This is almost 30% of the total amount of carbon currently added to the atmosphere each year. Clearly, if the potential absorption and sequestration estimated here could be realized, it would make a major contribution toward cleaning up the greenhouse problem.

But the estimate is open to question on two counts: one concerns the methods used to arrive at the estimate and the other concerns the absence of an analysis of the economic feasibility of the many measures that would have to be undertaken to actually achieve the estimated potential reductions.

The methods issue. Reclaiming salt-affected soils: the discussion of this source is in about two pages of text. The discussion deals with experimental work in reclaiming salt-affected soils, most of the experiments being in India. The actual estimate of the potential reduction in carbon emissions by rehabilitating these soils is in a single paragraph, referring to the experimental work in India.

Erosion control: the discussion of how erosion control reduces soil carbon emissions is on less than one page and some detail on how the actual estimate was made is in Table 3-8.

Restoration of degraded soils: Table 3-22 lists three aspects of soil restoration and the potential carbon capture possible from each. The three aspects are erosion control, restoration of strongly/extremely eroded soils, and adoption of recommended practices on slightly/moderately eroded soils. The inclusion of erosion control as an element in restoration of degraded soils is confusing. Comparison of Table 3-22 with Table 3-8 suggests that the Table 3-22 estimate of carbon reduction from erosion control was derived from the Table 3-8 estimate of the reduction from the same source. If so, this would seem to indicate significant double counting since the emission reduction attributable to erosion control in Table 3-22 is over 15 percent of the one billion ton reduction estimate for all sources. I suggest that the authors clarify what is going on here.

Carbon credit for biofuel production: The argument here is that planting trees, shrubs or grass on degraded land in dry areas and maintaining the increased biomass indefinitely will increase the supply of biofuel, which will substitute for fossil fuels, thus generating a carbon emission reduction credit for improved use of degraded land. The argument for the credit is put in a single paragraph.

In my judgment, the arguments underlying the estimates of potential carbon reduction from the four sources are so skimpy and the brevity of them suggests the reason for my judgment. But let me go a bit further with an example. As indicated above, Table 3-8 gives some detail on how the estimate of carbon emission reduction from control of soil erosion was done. A key detail is the assumed amount of erosion per hectare per year on slightly, moderately, strongly, and extremely eroded land in dry areas around the world. The assumed erosion rates are in relation to T values, which were originally established by the U.S. Department of Agriculture's Soil Conservation Service. T values are supposed to measure the maximum amount of erosion per hectare per year that a soil can sustain without suffering a long-term loss of productivity. Lal and colleagues assume, *without any supporting argument*, that per hectare erosion rates rise in an orderly way from 1.25 times T on slightly eroded soils to 3 times T on extremely eroded soils. The implied annual erosion rates go from 14 tons per hectare per year on slightly eroded land to 34 tons on extremely eroded land.

Rattan Lal is a world leader in discussions of the problems of estimating global amounts of soil erosion, and has often pointed out that in fact we have no scientifically well-founded estimates. Given the charge he had in preparing this paper, he had to make assumptions about how much erosion is occurring. If I had had the same charge I would have made the same sorts of assumptions; and if Lal were commenting on my effort, he would express the same skepticism about the outcome as I express here.

The economic analysis issue. Lal and associates are not economists and cannot be faulted for not discussing the economic feasibility of the various measures they propose to reduce carbon emissions from degraded land. That the authors recognize that their analysis is incomplete is clear from the fact that in most cases they refer to their estimates as *potential* reductions in soil carbon emissions.

Some of the economic issues raised by efforts to reduce soil carbon emissions are covered in another chapter, and I will not venture into that area in any detail here. However, a key question is how to provide farmers incentives to adopt the farming practices that will achieve the sorts of reductions in soil carbon emissions that Lal and colleagues have discussed. Research over the last three decades leaves no doubt that when farmers are given opportunities to profit from adoption of new practices, they will adopt. Two fundamental conditions define these opportunities: (1) the risk-discounted returns to the farmer of adoption must exceed the risk-discounted costs, and (2) the farmer must be confident that if he makes the investments needed for adoption he will reap the benefits.

Meeting the first condition is sometimes referred to as "getting the prices right." There is something, but not everything to this. The prices that the farmer pays must of course be "right" in relation to those he receives for his product. But for this to happen farmers must have reliable access to markets for both their outputs and their inputs, to credit at profitable rates, to flows of information about how to get the most out of the new practices, and, perhaps most important, they must live in a political and social environment that promises some stability from one day to the next as they go about their business. It is common knowledge that these conditions are often not met in the developing countries.

The key to the second condition is clear, enforceable property rights in the land and other resources farmers use in production. When that condition is met, farmers will be reasonably confident that they will reap the returns that will flow from the investments they must make to adopt the new practices.

At present, policy makers in the developing countries are not likely to be much impressed by arguments that they should promote changes in farm practices to offset soil carbon emissions. This could change should a regime emerge in which governments receive credits for carbon sequestered within their borders. Even without such a regime, however, policy makers who are serious about their responsibilities will be impressed by serious arguments for policies to provide farmers new opportunities to produce more while at the same time protecting and improving the natural resource base. Reductions in soil carbon emissions could be indirect, but, judging from the work of Lal and colleagues, nonetheless important consequence of such policies.

A Final Comment

Lal and colleagues were asked to confine their analysis to the carbon sequestration potential of better management of degraded soils in dry areas. Drawing on GLASOD data collected by the people at the Agricultural University of Wageningen, they show that there are 1,137 million hectares of such land around the world. However, the GLASOD data show that there are also a little over 800 million hectares of degraded land in humid areas (Oldeman, 1992). Because moisture is less limiting for this land, its production potential probably is substantially higher than that of degraded land in dry areas. Improved management of degraded land in humid areas therefore might not only have higher economic payoff, thus making it easier to sell to farmers, but it might also have relatively high potential for per acre sequestration of carbon in soil.

CHARLES F. HUTCHINSON, *NASA Headquarters, University of Arizona*

The topic addressed differs fundamentally from the others presented. Rather than dealing with basic issues in science, technology, and economics, this paper considers the challenges of how we might move forward and apply them to a complex set of problems on the ground. Thus, it must touch on topics covered more fully in other papers. Moreover, unlike other papers, this one deals with issues that tend to be found more often in developing countries.

Desertification presents a number of features that make it difficult to address in such a short treatment:

1. It is global, and occurs in many different physical, economic, and social environments.
2. It involves a number of different physical processes—all of which result in a decline in SOC.
3. It generally affects populations who are marginalized economically and politically.

The desertification topic has been a global issue for a quarter century, and has been addressed with little success (i.e., the total failure of national plans of action following the 1976 United Nations Conference on Desertification). This poor performance may be attributed to several factors. One is indifference to the problem at both the national and international levels. As noted above, affected areas tend to be marginal both economically and politically. Thus, they seldom draw sustained attention and rarely make national agendas. Internationally, there has been little interest in dryland development in general, and very little interest in reclaiming damaged areas. Although they have taken different action recently, in the mid-1980s the World Bank and the U.S. Agency for International Development (USAID) both adopted policies the focused on irrigation in drylands. They argued that this represented the best return on investment and thus largely ignored rainfed croplands and rangelands where the bulk of desertification problems occur (in terms of area affected). In addition to indifference, a number of misguided schemes were proposed over this period that were either economically or socially infeasible, or just plain silly (e.g., the recurring trans-Sahelian Green Belt proposal to contain the Sahara with trees).

The American Dust Bowl of the 1930s offers a model for dealing with desertification. Then, farmers in marginal areas were hit hard by drought, and the two factors combined to create catastrophic erosion in many places. However, programs of extension introduced new soil and water conservation technologies, and a system of subsidies helped to ensure that they were adopted and maintained. In addition to technology, this required an institutional capacity and money to put it in place. Of these three critical factors (i.e., technology, institutions, and money), we find only one in the Sahel. Technology to deal with the problem has been available for some time but as already noted it has not been widely adopted.

Soil carbon sequestration in the Kyoto mode promises something new. It addresses the foremost of these missing critical factors—money—and in so doing promises to force a solution to the third issue (institutions).

Soil carbon sequestration offers a program with a specific focus, a firm scientific footing, and a funding mechanism to address the intertwined problems of land degradation, poverty, and food insecurity. In the "win-win" context, the "wins" that might be gained in these regions could be of more human significance than any other.

Unresolved Issues

The paper addresses a wide range of issues, but most of them tend to revolve around the technology of soil carbon sequestration. This leaves many unresolved issues upon which success will ultimately hinge.

Equity: It is possible to consider the "win-win" situation in other senses. For example, in some situations land productivity is enhanced (an individual/group benefit) while C is sequestered (a societal/global benefit). On this point there seems to be little disagreement. However, as the authors point out, it does require that individual farmers make some initial investment (e.g., construction of stone bunds for water control) on which a return cannot be realized immediately. More specifically, the authors argue that extremely degraded lands be taken out of production and converted to production of "biofuels" that might be used for power generation. These would not come on-line for two to three years. Again, in these extremely poor areas, it is unlikely that farmers or pastoralists would be in a position to forego income from these lands for that period without outside interests sharing part of this burden.

Technology: The paper focuses almost exclusively on technical interventions. None of them is new and most have been widely promoted in the developing world for at least two decades. In view of the authors' proposals, this is not necessarily bad because the technologies have been well proven. There is cause for concern, though, because these same technologies have not been adopted as readily as might be expected.

It can be argued that part of the failure to adopt is related to larger economic and policy factors (see above). However, at the "micro" level, the failure to adopt also suggests that either the technologies do not perform as claimed or, more likely, that obstacles other than ignorance are restricting adoption. Some of these are undoubtedly economic (see above), but other factors may be at play (e.g., tenure insecurity, credit access, labor availability).

Regardless of the causes of "apparent" resistance to innovation, a parallel effort is required to understand what they might be and to design interventions that are sensitive to them. Typically, such interventions would be built on existing systems of agriculture and resource management to (a) minimize economic and social disruption and (b) enhance the likelihood of ready adoption.

Institutional Alternatives: There are arguments from the environmental side against schemes for carbon sequestration in general. Sequestration is only a stopgap that, without measures to dramatically reduce the rates of C emission, will only prolong the search for an ultimate solution. However, regardless of these assertions, it does seem that SOC sequestration is a "win-win" situation, as argued by the authors. If there are losers, it is unclear who they might be.

The "winning," however, goes beyond benefits suggested by the authors. Inherent in any carbon trading scheme is an element of equity in which some portion of the "goods" that come from C emissions is shared with those who must bear part of the environmental costs. The disparity between the developed world and those areas that experience desertification is profound, particularly in terms of income. Sequestration of SOC represents a mechanism to address inequities between these two poles of income, but also may offer the elusive key to desertification control: providing an economic incentive to farmers and herders to break the cycle of increasing environmental dependency and consequent degradation.

The issue of monitoring progress so that compensation might be distributed accordingly is addressed in passing by the authors. Technically, this may prove to be a challenge, but perhaps not a major obstacle. Institutionally (i.e., who will do what) the issues are not so clear, particularly when determining how the benefits will be distributed in the most practical sense. The mechanisms by which funds might be transferred to the agents of SOC sequestration (i.e., local farmers and herders) and how information would flow from monitoring to disbursement efforts is far from clear and not addressed by the authors. Given the poor performance or failure of so many political institutions in the developing world, this issue might be the most significant obstacle confronting a successful SOC sequestration campaign. However, the very existence of a global mechanism by which C might be traded could very well force the development of new, more responsive and accountable institutions that would better serve these impoverished populations.

ROB JACKSON, *Duke University*

That soil carbon storage in desertified lands is being considered as a way to slow the increase in atmospheric CO_2 speaks to the daunting extent of desertification globally. Deserts have substantially less organic carbon in their soils than do tropical and temperate forests, and the carbon stored in desert vegetation is trivial compared to that stored in forest vegetation. But with as many as two billion hectares of arid and semi-arid lands degraded (the figure quoted by Lal et al.), restoring a large portion of those lands would have the benefit not only of slowing the rise in atmospheric CO_2, but of improving the lives of millions, perhaps billions, of people worldwide.

The analysis of Lal et al. presents many possible tools for reversing desertification, and the diversity of such techniques is encouraging. The challenge is when, where, how, and sometimes, whether to implement those technologies over millions of hectares, the only scale relevant for storing carbon in a Kyoto context. This does not mean that applying the tools at smaller scales is not desirable, just that their broad scale application is the only way to store sufficient carbon to reduce atmospheric CO_2 significantly.

There are many reasons that we all hope the effort to restore degraded lands and increase soil carbon storage is successful.

The following discussion highlights six factors that might be considered before implementing the tools described by Lal et al. I highlight these topics as ones that were not discussed in the paper or that could be discussed more fully (recognizing that the chapter was already very detailed and thorough).

Economic Constraints on Soil Carbon Sequestration in Desertified Lands

The prospects for success of the many technologies presented by Lal et al. may need to be tempered by the economic limitations of applying them over millions of hectares. As desirable as the technologies may be, their application in arid lands may not return the initial investment for private landowners. Additional sources of revenue, perhaps associated with a carbon tax, may be needed to provide landowners with incentives for restoring their lands.

Different Strategies for Different Places

What strategies are appropriate and how they might be implemented for restoring degraded lands will depend on the country, the particular region of the country, and local conditions. Whether land is publicly or privately owned is one important factor. In the United States many of our degraded lands are public, administered by such agencies as the Bureau of Land Management and the National Forest Service. These public holdings are in some ways an advantage for implementing policies to restore degraded lands (ignoring the social difficulties of policy implementation), since policy changes can be administered unilaterally across broad areas. When the same area is owned by many private landowners, the policies and incentives must obviously be different.

Many other factors will also affect implementation of the technologies. As mentioned above, the economic status of the country or landowner will affect the likelihood of success. A second issue is population growth. Where populations are growing rapidly (as in many areas where desertification is rampant), it may be difficult or impossible to reverse land degradation. Finally, local factors, both social and scientific, must be incorporated into any plan for reversing desertification (Rayner and Malone, 1997); the people using the land must see the benefits of the policy or it will never be successfully applied.

The Interaction of Grazing and Desertification

The restoration and sustainable use of rangelands are needed for reversing one cause of desertification. For example, in the late 1800s the number of sheep in New South Wales, Australia was 13 million; ten years later the number dropped to <5 million, where it remains today (Perry, 1968). By restoring degraded lands, we should be able to manage them sustainably and productively, perhaps even at stocking rates above current levels.

The tools for grazing management are many, including controlling animal density, the timing of grazing, its duration, and the frequency of rotation. These tools should be applied creatively to rangelands worldwide. Economic and social incentives must also reward long-term stewardship of the land. A carbon tax and other economic incentives might allow some of the most marginal rangelands to be rested, temporarily and permanently; lost income from reduced grazing might be replaced by new income for storing soil carbon.

Interactions of Carbon Sequestration with Other Environmental Factors

Until very recently, and perhaps even now, environmental problems were often approached in isolation. If the salinization of a river occurred, constructing a desalinization plant was the solution, rather than treating the cause of the increased salinity. The lesson for soil carbon sequestration is that the tools and technologies discussed by Lal et al. need to be applied in ways that minimize other environmental problems. For example, if trees are to be planted for carbon sequestration or biofuel production, severely degraded lands should be planted first, rather than converting already vegetated lands. Applying the tools in ways that minimize other environmental problems may seem obvious, but predicting the consequences of broad-scale management is difficult at best.

The Importance of Soil Depth

Analyses of soil organic carbon have typically examined only the first meter of soil. Two recent analyses (Batjes, 1996; Jobbagy and Jackson, 1999) have begun to address deeper pools of soil organic carbon. Jobbagy and Jackson used two global soil databases—the World Inventory of Soil Emission Potential Database (WISE) compiled by the International Soil Reference and Information Centre (Batjes, 1996) and the National Soil Characterization Database of the United States

Department of Agriculture (USDA, 1994)—to examine organic carbon storage to three meters depth globally. The first meter of soil contained approximately 1600 Pg of organic carbon (similar to values found in other analyses), but the second meter contained > 500 Pg and the third meter contained almost 400 Pg (Jobbagy and Jackson, 1999). The analysis indicates that the second and third meters of soil alone contain more carbon than does the atmosphere. This carbon will turn over more slowly than that in the first meter of soil; if ways are found to increase relatively deep soil carbon it is likely to stay in the soil for centuries at least. Vegetation types with different allocation to roots and with different rooting depths may provide one way to accomplish this goal (Jackson et al., 1996; Canadell, et al. 1996).

In a biome analysis of the same dataset, we also showed that on a relative basis deserts have the highest proportion of soil organic carbon in the second and third meters of soil (86% of the amount of soil organic carbon contained in the first meter is contained in the second and third meters of deserts). The relative importance of deep soil carbon in deserts and its relatively slow turnover time may provide an opportunity for substantial soil carbon storage.

The Role of Vegetation Type in Soil Carbon Storage

Climate and soil texture have long been recognized for their important role in global patterns of soil organic carbon (e.g., Jenny, 1941; Post et al., 1982; Schlesinger, 1997). More recently, it has become clear that vegetation type (e.g., grasses, shrubs, and trees) also plays an important role (Jobbagy and Jackson, 1999). We used the National Soils Characterization Database of the USDA to analyze soil organic carbon content in the first meter of soil for U.S. deserts (< 300 mm precipitation; 5-20°C mean annual temperature). All soils having natural vegetation dominated by either grasses or shrubs were used in the analysis (n=67). While temperature and soil texture explained 42% of the variation in soil organic carbon, the addition of a dichotomous variable for grasses and shrubs improved the model fit an additional 12% (r^2=0.54, p=0.0002 for the additional vegetation variable in the multiple regression).

More importantly, shrub-dominated sites had only 60% on average of the soil organic carbon in the first meter as contained in grass-dominated sites. Since desertification frequently coincides with a loss of grasses and an increase in shrubs, restoring perennial grasses to degraded and desertified lands may help store substantial amounts of soil organic carbon, in addition to curbing erosion and other undesirable consequences.

References

Batjes, N.H. 1996. Total carbon and nitrogen in the soils of the world. *European Journal of Soil Science* 47:151-163.

Canadell, J., R.B. Jackson, J.R. Ehleringer, H.A. Mooney, O.E. Sala, and E-D. Schulze. 1996. Maximum rooting depth for vegetation types at the global scale. Oecologia 108:583-595.

Jackson, R.B., J. Canadell, J.R. Ehleringer, H.A. Mooney, O.E. Sala, and E-D. Schulze. 1996. A global analysis of root distributions for terrestrial biomes. *Oecologia* 108:389-411.

Jenny, H. 1941. *Factors of soil formation*. McGraw-Hill, New York.

Jobbagy, E.G. and R.B. Jackson. 1999. The vertical distribution of soil organic carbon and its relation to climate and vegetation. *Ecological Applications*, submitted.

Perry, R.A. 1968. Australia arid rangelands. *Ann. Arid Zone* 7:243-249.

Post, W.M., W.R. Emanuel, P.J. Zinke, and A.G. Stangenberger. 1982. Soil carbon pools and world life zones. *Nature* 298:156-159.

Rayner, S. and E. Malone. 1997. Zen and the art of climate maintenance. *Nature* 390:232-234.

Schlesinger, W.H. 1997. *Biogeochemistry: an analysis of global change*. Academic Press, San Diego, USA.

USDA. 1994. *National Soil Characterization Data*. Soil Survey Laboratory, National Soil Survey Center, Soil Conservation Service. Lincoln, Nebraska, U.S.A.

LARRY P. WILDING, *Texas A&M University*

The authors of this paper have done a laudable job in synthesizing a voluminous literature on a very difficult and elusive topic, desertification. The report contains a vast database on "global desertification" with major focus on the drylands of West and Central Africa and India. They have collated this information into a well-organized and informative treatise, despite the shortcomings of limited hard datasets with well-documented methodology over broad geographical regions sufficient to quantify C sequestration predictions. In this regard some of the datasets may be reported beyond warranted limits. The multiple facets of desertification controls on C sequestration elucidated in this position paper include definitions, extent, forms, processes, impacts, control strategies, land quality indicators, policy issues, data inadequacies, and research priorities. Most of the focus for increased C sequestration is on above- and below-ground biomass. Little attention is given to the potential for sequestration of inorganic C in such degraded environments.

They have nicely blended the balance for food security and other necessities of life, especially for the developing world, with environmental goals to control desertification. After all, enhanced natural resource protection, preservation, and restoration are consistent with holistic sustainable biosphere objectives and not restricted to any specific benefit accrued. Food security is commonly the number one priority issue of a developing world where risk avoidance has limited alternatives. On the other hand, in developed countries incentives to balance environmental protection against economic benefits may be the first priority. Thus, as so eloquently articulated by the authors, the case for enhanced natural resource protection and restoration that result in increased C sequestration must be a win-win game for all sectors of society politically, economically, and scientifically.

A few additional points on this generally well-written pivotal paper follow:

Scope of Desertification

This report identifies the extreme difficulty and complexity in defining desertification because of its multiple origins, processes, terminology, and bias towards specific geoclimatic regions of mid to low latitudes. Not surprisingly, much of the literature cited comes from soils in drylands of Africa and India. However, many soils in drylands of Africa and Australia are acid, iron-rich, low-activity clay systems that were derived from old, stable bedrock platforms and preweathered, polycycled sediments. Hence, the only reliable way to scale data from one locale to another, or to regional or global levels, is by using soil databases with appropriate classification identities that reflect similar soil attributes.

The central concept of desertification focused on drylands leaves out regions of the world that have similar degradation from anthropogenic activities. For example, high latitudes such as Iceland, steeplands such as in Central and South America and the Caribbean region, and saline-sodic or acid sulfate soils of humid coastal environments experience the same kind of degradation (desertification) effects as in the tropical and subtropical drylands. Hence, the regions of the world undergoing desertification may be considerably more extensive than reported in this chapter.

C Sequestration Projections

In this section the authors could expand on the rationale used for calculations and add appropriate references to justify their assumptions of potential C sequestration to be gained from control strategies. Landscape geometry becomes an important element in determining the global magnitude of desertification because landscapes are anistropic laterally and vertically where relief plays a major role in water balance, salt balance, erosion, intrinsic soil properties, and the propensity for soil desertification. Wind and water eroded sediments, while enriching receiving areas in nutrients and more favorable land quality indicators, constitute a double-edged sword because source areas have been depleted in these same attributes.

Caution needs to be exercised in interpreting enhanced pools of C sequestered by the array of desertification control options presented. In addition to the uncertainties in scaling up the soft database to global levels and inability to place confidence limits on the estimates, many of the root causes for soil erosion, physical and chemical degraded lands, and salt-affected soils are interdependent. Predictions of enhanced C sequestration by adopting amelioration strategies and/or removal of land from production may result in "double credits." For example, physical and chemical degradation—crusting, compaction, soil infertility, soil toxicity, and subsoil restrictive layers—are common causes for soil erosion.

Role of Inorganic C in C Sequestration

The role of soil inorganic C (SIC) in C sequestration is less well understood than that of soil organic C (SOC). Based on recent research in our laboratory, inorganic

C sequestration over the period of geologic development of the soils studied is up to 25 times greater than SOC (Wilding et al., unpublished data). SIC may be a sink, source, or have no net effect on atmospheric CO_2. Pedogenic carbonates formed from base-rich bedrocks or noncarbonate sediments are a sink. However, pedogenic carbonates in calcareous parent materials are neither sink nor source, because for each mole of CO_2 sequestered a mole of CO_2 is released upon precipitation of the calcium carbonate.

In systems of partial or complete soil leaching, the major mechanism for sequestering SIC is via movement of HCO_3 into groundwaters or close systems that have limited exchange with ambient environments. Based on reconstruction of carbonate fluxes in soils formed in strongly calcareous parent materials over geological time periods, this could account for 0.25 to 1.0 ton SIC/ha/yr. While this mechanism may appear to be of limited impact for degraded drylands, certainly it has implications when groundwaters undersaturated with $Ca(HCO_3)_2$ are used for irrigation. Enhanced primary productivity of the vegetation and salinity control strategies (e.g., gypsum amendments, organic wastes) can result in increased leaching of $Ca(HCO_3)_2$ under irrigation if the irrigation waters are not already saturated with respect to bicarbonates.

Dissolution of exposed carbonates in soil systems by acid rain, nitrogenous fertilizers, oxidation of iron sulfides, or organic acids may result in a source of greenhouse gases if subsequent precipitation of pedogenic carbonates occurs. However, if the dissolved bicarbonates are either leaching through the soil or removed by overland flow without subsequent precipitation of carbonates, then this mechanism is a transient sink of intermediate to long residency rather than a source of atmospheric CO_2.

Desertification Control Strategies

Many of the strategies for C sequestration, restoration of severely degraded lands, and soil/water/vegetative management protocols are generic, lacking design specifics, quantification, evidence of data uncertainties, and probabilities of success. The authors note these limitations in their introductory statement.

Biological control strategies and biodiversity impacts are nicely captioned in the report. However, the opportunities for breakthroughs via biotechnology are underemphasized. Here is an excellent opportunity for soil scientists to contribute their expertise to help biologically design plant species that better accommodate adverse, desertified environments. Biological control strategies need to focus on deep-rooted, drought-tolerant, toxicity-resistant species that can be coupled with other biological or physical intervention control strategies. Traditional soil and water conservation strategies, such as conservation tillage and surface mulching, may have limited value in steeplands where major erosion is by block slumping rather than sheet and rill erosion. Shallow-rooted species and conservation strategies that ignore these landform processes are ineffective in tying thin soils to substrata bedrock for stability on slope gradients of 40 to 75%. Vetiver (*Vetiveria spp.*) and a number of the leguminous tree species cited by the authors in this report are useful in this regard.

Future Research Priorities

To research priorities listed in the report the following can be added:

- Develop a means to identify early warning symptoms of desertification and land quality degradation via biological (microbiological) indicators
- Quantify the role of SIC in C sequestration processes, rates, dynamics, and magnitudes
- Develop plant species that tolerate adverse physically and chemically degraded soil conditions; enhance C sequestration; and provide for food, feed, fiber, and fuel supplies
- Enhance efficiency of irrigation systems via precision agriculture technology to adjust application of dosing rates to fit variable physical and chemical conditions of soil systems.

The above commentary is not intended to detract from the generally high quality of the paper nor the excellence of the individuals who provided this summary on desertification control to sequester C and mitigate the greenhouse effect. It was an arduous task and one accomplished most successfully.

4

Soil Carbon: Policy and Economics

Gregg Marland, *Oak Ridge National Laboratory*

Bruce A. McCarl, *Texas A&M University*

Uwe Schneider, *Texas A&M University*

Within the context of the United Nations Framework Convention on Climate Change (1992), the Kyoto Protocol (1998), and subsequent discussions, a number of features make it attractive to consider the prospect of trying to mitigate increases in the atmospheric concentration of greenhouse gases through increased carbon sequestration on agricultural lands (defined broadly to encompass forestry lands). Certainly the Kyoto Protocol holds open the possibility that interests outside of the agricultural sector may approach those in the agricultural sector in a market setting that permits emissions trading and credits for carbon sequestration. Subsequent transactions would involve the purchase of emissions rights (from those with the ability to sequester carbon) so that emitters with high costs of emission reduction could pay others to reduce net greenhouse gas emissions in lieu of substantially reducing emissions themselves.

The Kyoto Protocol explicitly mentions forestry as one of sinks of greenhouse gases for which sequestration credits can be claimed, and there is further wording in the Protocol that might permit other agricultural sinks to be developed over time. Furthermore, it is well understood that soils are one of the planet's largest sinks for carbon and hold potential for expanded carbon sequestration through changes in management.

This (the St. Michaels) Workshop primarily has considered the overall question of carbon sequestration in agricultural soils. This paper considers the policy and economic dimensions of the question, providing broad coverage of the overall issue. In structuring this paper we are not trying to present conclusive, completely crafted arguments but rather a system of points to stimulate thinking.

We organize the paper around 14 themes.

1. The Kyoto Protocol and the context for managing agricultural soil carbon

2. Agricultural practices to be encouraged

3. Positive externalities
4. Negative externalities
5. Longevity of agricultural carbon and keeping carbon that has been paid for
6. Monitoring, verification, and soil carbon sequestration
7. Private property rights
8. Energy policy and carbon sequestration
9. Farm income supports and carbon sequestration
10. Practical economics
11. Farmer behavioral impediments and enhancements
12. International issues
13. Institutional and social issues
14. Soil carbon as part of a more complete agricultural response agenda

THEME 1
The Kyoto Protocol and the Context for Managing Agricultural Soil Carbon

The United Nations Framework Convention on Climate Change (UNFCCC) defines as its objective the "stabilization of greenhouse gas concentrations in the atmosphere at a level that would prevent dangerous anthropogenic interference with the climate system." In order to give substance to this objective, the Parties to the UNFCCC met in Kyoto, Japan, in December 1997, and drafted a Protocol that they intended would provide binding limits on greenhouse gas emissions and begin the process of limiting their atmospheric concentrations. Most observers (e.g. Bolin, 1998: Wigley, 1998) recognize that, if ratified, the Kyoto Protocol will provide only the beginning step in this process.

Although the prospect of implementing the Kyoto Protocol raises many issues, the Protocol does establish international concurrence on some basic ideas and principles. For example, the Protocol recognizes that its objective of stabilizing the concentration of atmospheric CO_2 can be met either by decreasing the rate at which CO_2 is emitted to the atmosphere or by increasing the rate at which CO_2 is removed from the atmosphere. Our focus here is on carbon flows to and from soils and the political and economic setting for enhancing carbon storage in soils. Even without ratification, the Kyoto Protocol helps to define this setting.

The decrease of carbon stocks in the biosphere has historically been a net source of CO_2 emissions to the atmosphere, but this process is amenable to reversal. Net carbon flows from the atmosphere to the biosphere are feasible. The Kyoto Protocol clearly establishes the concept of credits for carbon sinks (Article 3.3). However, three concerns about carbon sinks are evident in the Protocol and supporting papers. First, negotiators wanted to limit credits to activities undertaken purposefully—e.g., there would be no windfall credits for carbon taken up as a result of CO_2 fertilization (the increased rate of plant growth because of increased CO_2 in the atmosphere). Second, negotiators sought to limit credits to those that could be accurately and reliably measured. Third, negotiators did not want to allow countries to meet their full commitments by increasing biospheric sinks. Consequently the Protocol permits credits for a limited list of activities (Article 3.3) while providing the option that other activities could be added to this list later (Article 3.4).

The Kyoto Protocol limits credits for carbon sinks to afforestation and reforestation since 1990. Forest soil is part of the forest so that, to the extent that credits can be taken for afforestation and reforestation activities, the increasing carbon storage in forest soils can (presumably) produce credits. This will be limited to carbon sequestration that can be verified. At this point these credits are not guaranteed, however. The Protocol fails to define "forest," "afforestation," and "reforestation"; the Parties to the Protocol are likely to endorse specific definitions for these terms to establish which carbon reservoirs are intended to be included. The Intergovernmental Panel on Climate Change has been charged to prepare a special report that will explore, among other matters, the technical issues raised by alternative definitions of these terms.

Agricultural soils are specifically mentioned in Article 3.4 of the Kyoto Protocol: the Conference of the Parties will "decide upon modalities, rules and guidelines as to how, and which, additional human-induced activities related to changes in greenhouse gas emissions by sources and removals by sinks in the agricultural soils and the land-use change and forestry categories shall be added to, or subtracted from, the assigned amounts..." Simply stated, sequestration in agricultural soils is not now permitted to produce carbon sequestration credits under the Kyoto Protocol, but the door is left open for its addition.

A final item of importance is that carbon sources and sinks in the biosphere are to be "measured as verifiable changes in carbon stocks." This is of particular interest in the case of harvested wood products, but it does reinforce the principle that sustainably produced biomass fuels will not be counted as a source of CO_2 emissions, nor will the producing forest be credited as a continuing sink for carbon. Other greenhouse gas emissions from the agricultural sector (e.g., methane emissions from enteric fermentation, manure management, rice cultivation, and field burning; and N_2O emissions from soils) would be counted under the Kyoto Protocol at the time and place that they actually occur.

THEME 2
Agricultural Practices to Be Encouraged

If credit toward national emissions targets can be gained by increasing the stores of carbon on agricultural and forest lands, various land-management practices might be encouraged. This is the subject of another paper in this series (Metting et al., this volume), but we list a sample of such practices for the sake of illustration (see also Lal et al., 1998). These practices include some that would increase the amount of carbon stored, either above and/or below ground, and others that would decrease the loss of carbon from the biosphere. Note, however, that of the items listed below, only forested acreage increases and better management of reforested areas appear eligible for emission-reduction credits under the current phrasing of the Kyoto Protocol. Article 3.4 of the Protocol leaves the way open to add any or all of the other items in the list at some future time.

- Reduced tillage
- Longer forest rotations
- Increased forested acreage with conversion of land from agricultural uses
- Improved forest management of existing forests
- More carbon friendly forest harvest practices
- Better management of reforested areas
- Restoration of degraded lands
- Retirement of agricultural lands into permanent grass cover—Conservation Reserve
- Management of residues in forest and agricultural harvests.

Land management decisions affect a larger physical system than the landscape itself. There is, for example, a trade-off between the amount of carbon stored on forest lands and the production of forest products. Marland and Schlamadinger (1997) pose the question whether the net carbon emissions are greater when trees are grown to sequester carbon or when trees are grown to displace fossil fuels and energy-intensive products. Modeling exercises suggest that when forest land has high productivity and when products are produced and harvested efficiently, net emissions are often less when land is used to produce products than when it is used to store carbon. This may be especially relevant in a political context where a limited range of activities can produce carbon sequestration credits, but displacement of fossil fuel use always yields credits.

THEME 3
Positive Externalities

Externality is a term that economists commonly use to describe the effects that policy designed to achieve one aim has on the welfare of individuals not directly associated with that aim. An externality might, for example, include an alteration in realized soil erosion under a policy designed to increase carbon sequestration. Here we discuss positive externalities that might occur in environmental and economic accounts when policy to promote enhancement of carbon sequestration on agricultural lands is implemented.

Pursuit of carbon sequestration policies can have a number of beneficial and possibly unintended effects. McCarl et al. (1997) examined the effects of higher energy prices caused by a Kyoto-Protocol-motivated carbon cap and trade system. They found that higher energy prices stimulated widespread expansion of area in reduced tillage. In turn this resulted in a reduction in soil erosion. The rate of adoption of reduced tillage and the rate of erosion reduction were both large relative to many of the policies and programs that have been developed under the U.S. Department of Agriculture (USDA) farm program conservation titles or other USDA Natural Resource Conservation Service programs. A country bears various costs because of erosion: reduced water quality, sedimentation, etc. Increased conservation tillage would reduce these costs. Thus, a policy based on carbon emissions or sequestration might well benefit a number of erosion-related areas not originally the target of the policy.

Other types of positive externalities could occur:

- Reduced tillage could alter soil organic matter, increasing soil water-holding capacity, leading to a reduced need for irrigation water.
- Expanded conversion of agricultural lands to grasslands or forests could stimulate wildlife populations.
- Reduced soil disturbance, and possibly diminished use of fertilizer, could alter the volume and chemical content of runoff from agricultural lands. In turn this would reduce water pollution, and improve water quality and the general ecology of streams, rivers, lakes and aquifers. Such alterations might improve the characteristics of the waters in these regions for use by non-agricultural water consumers.
- Diversion of agricultural lands into energy production to reduce CO_2 emissions might induce technological improvement in agricultural crops, permitting expanded electricity generation at lower cost.

Many other possibilities can be cited. Positive environmental and economic benefits (externalities) could arise out of policies intended to reduce CO_2 accumulation in the atmosphere.

THEME 4
Negative Externalities

Along with the possibility of positive externalities, where society reaps benefits in areas that are not directly the aim of carbon-sequestration-related policies, comes the possibility of negative externalities, where society encounters costs in areas that are not initially expected. For example, Adams et al. (1992) and McCarl (1998) show that programs designed to move agricultural lands into forestry could have deleterious effects on the traditional forest sector, leading to either deforestation of traditional parcels or reduced levels of management and lessened carbon sequestration that offset much of the total amount of carbon being sequestered.

Here is a short list of other possible negative externalities:

- Conversion of croplands to grasslands tends to increase emissions of the greenhouse gas nitrous oxide, but it also tends to increase the soil sink for another greenhouse gas, methane (Mosier et al., 1997).

- Reductions in intensity of tillage, coupled with the increased amount of plant material left on the soil surface, have been commonly found to require additional use of pesticides for weed, fungus, and insect management. This may have deleterious effects on ecological systems, runoff, and water quality.
- Expanded use of agricultural lands for carbon sequestration could compete with the use of agricultural lands for traditional food and fiber production. The result might well be decreased food and fiber production; increased consumer prices for crops, meat and fiber; and decreased export earnings from agriculture.

Again, many other cases could be cited. Negative environmental and economic benefits (externalities) could arise out of policies intended to reduce emissions or increase carbon sequestration.

THEME 5
Longevity of Agricultural Carbon and Keeping Carbon that Has Been Paid For

The soil is not an infinite sink for sequestration of atmospheric carbon. Instead, the soil absorbs carbon until it reaches a new steady-state equilibrium that reflects the new management environment. As the total level of soil carbon increases, net soil absorption of carbon decreases and the potential of the soil to become a net source of emissions increases. Thus, sequestration of carbon in soil can decrease the CO_2 concentration in the atmosphere only for a limited amount of time. Furthermore, changes in tillage practice in the future could cause the added carbon to be quickly released again.

The volatile nature of soil carbon and the possibility that future land-use changes could substantially alter the mass sequestered raises a policy design issue. In particular, suppose we subsidize reduced tillage, but the subsidy expires at a later time. Farmers may choose to go back to conventional tillage, reducing or eliminating the carbon gains. A big question is: how do we design policies that get the carbon into sequestration initially and then keep it there? This is also an issue with respect to forestry. A recent study by McCarl (1998) concluded that land converted from agriculture to forest under a carbon-based subsidy program would revert back to agriculture after one forest rotation, unless the program were somehow designed to prohibit harvest or to require that the land remain forested.

If sequestered carbon becomes a commodity that can be saved or sold, there would presumably be a system of credits and debits. Credits could be gained for carbon sequestered but there must be subsequent debits if the carbon is later released. In this sort of framework, a land owner could sell emission credits when carbon is sequestered, but would then be bound to either provide long-term stewardship for the sequestered carbon or to bear the cost of emissions when the carbon is released. The Kyoto Protocol has led to characterization of some lands as "Kyoto Forests" when they meet the qualifying criteria and are credited for sequestered carbon. In the same vein one can envision "Kyoto Farms" where carbon is stored in soils under some long-term obligation. A similar, but different, set of concerns arises if the "Kyoto Farm" is in a country that does not have emissions limits, and we will return to this under Theme 12 below.

Conversion of land from agriculture to forest raises yet another issue in terms of the mass of carbon sequestered in the traditional forest sector. McCarl (1998) found a substantial countervailing movement of land from the traditional forest sector back into the agricultural sector when a carbon subsidy caused large amounts of land to be moved from the agricultural sector to the forest sector. The issue, called leakage, points out the distinction between project-level issues and the issues involved with national commitments. Presumably this example would create no net effect, over time, at the national level, where one land owner would show carbon sequestration as a result of reforestation while the other showed carbon emissions as a result of deforestation. At the project level, however, the challenge is to account for this leakage and to distribute credits and debits appropriately. Again, there is a greater problem if these two land owners are in different countries, especially if the deforestation is occurring in a country without national emissions limits. There is also a time-dependency problem, because reforestation of a parcel of land takes up much less carbon in a given year than would be released by deforestation of a similarly sized parcel.

The tendency toward these offsetting strategies will depend on the details of the tax, regulatory, or incentive schemes that countries adopt to try to meet national objectives. And we need to recognize that an instrument like the Kyoto Protocol does not treat all carbon flows identically. A literal interpretation of the Protocol text would require that national accounts capture the countervailing emissions in forestry land out-migration above, but would not require reporting of the emissions if policies caused, for example, conservation tillage practices to be discontinued or grasslands plowed.

The current possibility for expanding agriculture as a carbon sink, but the risky or volatile nature of future releases, raises yet another policy possibility. Agriculture, with the possibilities for change in tillage practice and/or forest area, offers a near-term solution that may or may not yield carbon that is oxidized and released at a future date. However, an essential question is whether this is a way of reducing costs for current compliance while awaiting major technological breakthroughs that would substantially reduce compliance costs for future greenhouse gas emission reductions. Agriculture may provide a low-cost option bridging to an even lower cost set of future options, even if the sequestered carbon is lost in 20 to 40 years. This raises the possibility of futures speculation in carbon emissions credits, as land owners could sequester carbon and sell emissions permits when the price was high and then buy back these permits if the price drops at some later date.

There is also the risk that the ability to hold carbon in soils may diminish as the climate warms, because there is a negative relationship between higher temperatures and the organic matter content of soils. However, this may be offset by the consequences of increased plant growth.

THEME 6
Monitoring, Verification, and Soil Carbon Sequestration

A recurring theme in the Kyoto Protocol is the monitoring and verification of carbon emissions and sinks. A particularly important issue with respect to potential sinks is that they somehow be internationally certified. One needs to be able to monitor changes in carbon stocks. We will not discuss verifiability at length, as it is the subject of another paper in this volume (Post et al., this volume). What is of particular importance here is that the Kyoto Protocol includes a variety of flexibility mechanisms that permit, and would surely stimulate, trading in carbon emissions permits. Carbon sequestered by one Party could be used to offset emissions in another sector of the national economy (and thus help to meet national commitments under the Kyoto Protocol) or it could be traded or sold to a Party in another country to use in fulfilling its national commitments. In order to have a viable market in carbon credits, there needs to be a commodity that can be clearly identified and reliably and consistently measured.

We note also that there is a possibility that the value of carbon emission credits could depend on the uncertainty in their measurement. For example, Canada (1998) has outlined a proposal in which mitigation measures would report the amount of carbon sequestered along with the uncertainty in this measurement. Credits could be claimed only to the extent that there was 95% certainty in the amount of carbon sequestered. Under this procedure, the greater the uncertainty, the lower the mitigation credit that could be claimed in order to be 95% confident that the achievement was indeed at or above the credited amount.[1]

Another issue follows on the points developed in the last section. We may need to be able to show that the adoption of certain sink strategies does not lead to a substantial leakage from altered economic activity in other parts of the economy. For example, a policy for carbon sequestration that results in transfer of land from agriculture to forestry may stimulate substantial countervailing land transfers out of forestry, thus offsetting the carbon gains. Within Annex I countries[2] this may have a major impact on mitigation projects and the economics thereof, but little impact on national commitments. Even in Joint Implementation projects (see theme 12 below), or carbon trading activities between Annex I countries, this would have no impact on the total allocation of permitted emissions because emissions in one project or country would be balanced by comparable reductions in another project or country. However, if this kind of leakage occurs when one of the parties is a non-Annex I country, without emissions commitments, the result could be an uncompensated increase in the total global emissions allocation. That is,

[1]For example, if the measured value is +/- 5 (with 95% confidence), the target would have to be set at 50 to be 95% confident that the result was at least 45, but the target would have to be at 50 to be 95% confident of reaching 40 if the measured value is +/- 10.

[2]The Framework Convention on Climate Change lists in Annex I those developed countries and countries with economies in transition that should adopt practices to limit emissions of greenhouse gases. Countries not specifically listed in Annex I, developing countries, have no such commitments. Appendix B of the Kyoto Protocol lists many of the same countries and provides quantitative limits for greenhouse gas emissions.

there would be increased emissions permits for the Annex I partner without necessarily having compensating reductions by the non-Annex I partner. The problem is also severe at the project level, as one party may be accumulating credits when there is in fact little net gain. And, again, the problem is exacerbated because some changes in carbon stocks are reportable under the Kyoto Protocol but others are not.

And, finally, this raises the issue of baselines. Most notably under Article 12, the Kyoto Protocol requires that emissions credits arise from "reductions in emissions that are additional to any that would occur in the absence of certified project activity." This requires that we not only monitor and verify the carbon that has been sequestered but that we measure a baseline of carbon that would have been sequestered in the absence of a specific project. Measurement of carbon stock changes along the path-not-traveled might rely on modeling or control plots. In either case we have to distinguish between what did happen and what would likely have happened. Baselines are likely to be an issue under Article 3.4 of the Protocol as well.

THEME 7
Private Property Rights

As argued in McCarl (1998), embarking on the road toward enhancing carbon sequestration poses challenging policy questions regarding private property and could lead to conflict with private property rights. Adoption of more carbon-friendly systems in agriculture is not useful if offset by movement toward less carbon friendly systems on other lands or in future times. For example:

- If we start converting land from agriculture into forestry, we may need to ensure that these movements are not offset by countervailing movements to agriculture from traditional forestry.
- If we make alterations in grassland stocks through, for example, an expansion of conservation reserve program area, we may need to ensure that pre-existing grass and wetlands are not broken out into tilled uses.
- If we pay some farmers to adopt improved conservation tillage, we may need to ensure that other farmers, who are not being paid, do not revert to more intensive tillage.
- If we pay some farmers now to do certain things that expand carbon sequestration, we may need to ensure that they do not discontinue those practices in the future.

This means that total program design should consider features that discourage countervailing behavior that offsets current sequestration gains or future decisions to adopt behaviors that are inconsistent with carbon sequestration.

The design of an incentive scheme may pose particularly challenging policy questions and could encounter unintended expenses. For example, incentives designed to keep land in forestry might end up paying land owners who had no real intention of ever moving land out of forestry. On the other hand, if the program taxes or regulates alterations of land use, or the absolute mass of carbon emissions, then a private property rights issue will likely arise. This has been observed, for example, when land-use changes have been restricted in order to preserve endangered species whose habitat is dependent on private property. Will we allow existing forest owners, who are not being compensated in the program, to choose to deforest their lands and move them into agriculture? Will they be somehow taxed in proportion to any carbon released? How about existing farmers who are currently using some form of reduced tillage but later reverse that decision and use more intensive tillage systems? How about land owners who now have land under some form of grass or forest and convert it to tilled agriculture? How about land that is currently rather minimally disturbed in the agriculture or forest sectors but is converted into residential and/or urban uses that diminish the carbon storage potential?

All of these appear to be major property rights issues. Once we start creating restrictions or commodity value on carbon storage in agricultural land, where does it lead? This may be particularly cumbersome in a regime where treaty commitments are at the national level but all carbon flows are not accounted similarly.

THEME 8
Energy Policy and Carbon Sequestration

Rather profound changes in net greenhouse gas emissions are possible by using agricultural commodities as feedstocks for electrical power generation or liquid fuel production. In particular, if we use agricultural commodities to fuel power plants, then we offset use of mostly coal or natural gas. If we use agricultural commodities to produce liquid fuels, then we offset petroleum use. In either case the carbon released from long-term storage in the earth is replaced by a recycling of carbon via a renewable fuel, using solar energy and the photosynthetic process. The big question is the extent to which credits for reduced carbon emissions affect energy policy and energy prices. McCarl and Schneider (1999) show that displacing fossil fuels is expensive relative to other agricultural possibilities for carbon sequestration. Marland and Schlamadinger (1997) show that cross-sectoral impacts may make it difficult to properly attribute credits for carbon emissions reductions. As an example, increased use of wood in construction may increase carbon emissions from the forest products industry, in spite of larger savings through reduced use of fossil fuels in the concrete block or steel industries. A broad-based carbon tax might distribute the rewards for carbon sequestration activities differently than would other implementation strategies.

THEME 9
Farm Income Supports and Carbon Sequestration

Historically, agriculture has received substantial public subsidies in the form of price and income supports. These subsidies have made up a substantial portion of agricultural income. At times in the late 1980s or early 1990s the total cost of the farm program was one-half or more of estimated total U.S. net farm income. We spent $10-15 billion in the United States for a number of years on the agricultural farm program. Today we are in the process of phasing out the agricultural farm program and farm subsidies have been reduced to $4 billion or so, not counting the Conservation Reserve Program. But there is increasing pressure from farm interests to get back into the farm program business, particularly given low current prices for agricultural commodities.

From an economist's viewpoint several issues have contributed to the continuation of public subsidies to agriculture. Many say that we have the farm subsidies because we have a farm problem and need to preserve this industry. Several forces have influenced changes in agricultural incomes. Technological advances have caused the capability for the supply of food to grow faster than the total demand for food, and this has resulted in declining agricultural prices and farm income. A substantial portion of this technology-induced growth has been publicly funded, and perhaps the public should subsidize agricultural producers to make up for the effects of the technological shifts and income losses caused by this research funding. Today, in real terms, total income in the agricultural sector is somewhat less than 80 percent of what it was 25 years ago. Consumer food prices remain low even in the face of growing population. This has caused out-migration from agriculture and concern over whether we should subsidize producers to stay in the agricultural business. In addition, food security issues have been used to justify agricultural subsidies.

The greenhouse gas sequestration aspects of the Kyoto Protocol raise interesting new possibilities for income supports in agriculture. Most agricultural production faces what economists call an inelastic demand curve. People will not eat a great deal more even if food costs less, so increased production is often matched by declining prices. However, producing feedstocks for the energy market would probably place agriculture as a fairly small player in a market with what economists call an elastic demand curve. This market would not yield such large price reductions when agricultural production of energy crops is expanded. In this case production would tend to favor producers, yielding net farm-income benefits, as opposed to the consumer benefits that have arisen under the technological programs up to now. This could provide justification for a new breed of farm program, with funding based on energy benefits and carbon sequestration.

Even without income supports, improved markets for fuel feedstocks and carbon emissions permits could provide a new source of farm income. A problem with accepting payment for carbon sequestered is that, as pointed out above, carbon sequestration cannot continue forever because soil carbon will approach a new steady state. As carbon accumulates, the landowner accumulates an increasing responsibility for stewardship and, potentially, increasing financial liability if the carbon is subsequently released.

THEME 10
Practical Economics

From a practical standpoint there are a number of economic questions about carbon sequestration that need to be raised in terms of public policy and expected adoption. For example, are the comparative costs of carbon sequestration in agriculture low enough that they would justify inclusion of agriculture in an emissions trading market where non-agricultural interests could buy carbon sequestration credits from agricultural interests? Anecdotal evidence seems to suggest that this is the case, but the supply curve for the cost of non-agricultural carbon credits is not clear. McCarl (1998) and McCarl and Schneider (1999), among others, computed the cost of several agricultural opportunities to be somewhere at or below $30 per ton of carbon. Recent studies by The President's Council of Economic Advisors (1998), the Energy Information Administration of DOE (1998), and economists such as Manne and Richels (1998) have produced a wide range of numbers for the cost of carbon emissions reductions in other sectors. The range of costs depends very much on the trading regime permitted, i.e., the extent to which emissions credits will be traded internationally and which countries will participate. It also depends on the rate at which policies are implemented. If sequestration in agricultural soils provides low-cost carbon emissions credits, then there certainly will be room for a market where incentives large enough to stimulate behavioral changes in agriculture will be offered to producers.

We need to consider the aggregate price and production effects that a market in carbon emissions credits would have across the whole agricultural sector, and not just look at the cost of complying with something like the Kyoto Protocol. In earlier analyses of the economics of agricultural feedstocks for energy production, one of the major issues uncovered (Tyner et al., 1979) was that program adoption caused increasing prices of food to traditional consumers and reductions in export earnings to farmers. Crops for food and crops for mitigation strategies are competitors. Expanding crops for greenhouse gas emission mitigation would reduce crops for food. Food and land prices would likely rise. We need to account for all the costs and benefits of alterations in the traditional agricultural sector versus the costs and benefits of using land to meet obligations under the Kyoto Protocol. The total issue portends shifts in the distribution of income between agricultural producers and consumers.

We need also to consider the costs and benefits of the negative and positive program externalities, including, ideally, the costs and benefits of a changing climate.

THEME 11
Farmer Behavioral Impediments and Enhancements

Farmers do not seek only to maximize profits. A number of other factors will enter into their decisions as to whether or not to adopt practices that could increase the carbon content of soil. In particular:

- Many farmers are motivated by a stewardship role in terms of the soil and the environment. In that context one may find that farmers would more easily adopt soil-conserving techniques than would otherwise be the case.

- Risk often increases when less tillage-intensive practices are adopted in the place of conventional tillage. Thus farmers who are highly risk averse may choose to avoid adopting practices that would increase carbon sequestration.

This may be particularly true for older farmers who are not willing to bear the risk of changing management practices toward the end of their careers.

- Management requirements are likely to be more demanding when less tillage-intensive practices are adopted. Farmers may again be unwilling to adopt practices that require substantially more critical management activities. This again may be more characteristic of older farmers.

- Tillage practices, once adopted, have to remain in use for a long time if carbon sequestration gains are to be captured and maintained. Farmers may be unwilling to take on such long-term liabilities, and it may be difficult to pass the liability on from farmer to farmer when farm ownership changes. Lease terms for farmlands may also be problematic.

THEME 12
International Issues

Although most of the details are still to be resolved in international negotiations, the Kyoto Protocol establishes the principle of trading carbon emissions permits between countries. Article 6 discusses Joint Implementation, whereby one Annex I country can pursue projects in another Annex I country and use the carbon benefits toward its own emissions commitments. Article 12 discusses the Clean Development Mechanism, which provides similarly for projects with an Annex I sponsor and a non-Annex I host. Article 17 provides that emissions credits can be traded among Annex I countries. These various mechanisms do not necessarily have the same restrictions, however. Article 6 specifically mentions credits for "removals by sinks," but this language is missing from Article 12 and Annex I countries may or may not be able to use emission reduction credits from carbon sequestration in non-Annex I countries to contribute to their compliance with the Protocol.

Transfer of emissions credits under either Article 6 or Article 17 would not result in creation of any additional global emissions permits. What appears in the accounts of one party must be subtracted from the accounts of the other. Transfer of emissions credits under Article 12 would, however, result in creation of additional global emissions permits. Under Article 12 a country with emissions restrictions would obtain additional emissions permits from a country without emissions restrictions. Another issue still open to discussion with respect to international trading of credits is the allocation of responsibility for failure to adequately certify that carbon has indeed been sequestered. Does responsibility fall to the buyer or the seller? Sequestration in a non-Annex I country under Article 12 would raise additional questions. Presumably this sequestration would be undertaken under a defined contract and the sponsoring country would receive emissions permits as carbon was sequestered in the host country. After expiration of the contract, however, the host country would have no emissions limits under the Kyoto Protocol and the question is whether there is any liability if the sequestered carbon is subsequently released.

There are a number of related concerns about international leakage under a system in which some, but not all, countries participate. Some examples are described under Theme 7 above.

THEME 13
Institutional and Social Issues

A large number of institutional issues will be confronted in dealing with the opportunities and risks of carbon sequestration and trading of emissions credits. These range from the creation of formal markets for emissions credits to insurance against loss and mechanisms to aggregate markets to reduce risk. Institutions will be required to deal with certification and compliance and with legal and equity issues.

To be effective, an international agreement like the Kyoto Protocol needs some mechanism to maintain faith in its execution and to confirm compliance. Parties need to know that other Parties are acting in good faith and that their own claims are subject to confirmation. There is considerable international discussion underway about organizations and mechanisms to document and certify emissions reductions. There are both government and private proposals for third-party organizations, but substantial negotiation remains to be done.

There is also internationally expressed concern about issues of "environmental imperialism" and buying one's way out of national commitments. Are international mechanisms needed to guard the interests of Parties with fewer financial resources or poorer access to information? One can argue that trading of carbon emissions permits will occur in a free and open market, with transactions between willing partners; but there may need to be mechanisms to ensure that economic interests do not overrun social interests or equity issues. This may be especially true for issues of carbon sequestration because of the long-term commitments and implications involved.

Widespread adoption of different management practices in agriculture and forestry, especially practices that may require a higher level of management skills, will require a system of education and extension – an efficient network for technology support and technology transfer.

THEME 14
Soil Carbon as Part of a More Complete Agricultural Response Agenda

There are five principal ways in which agriculture might be affected by greenhouse gas mitigation efforts and emission trading markets:

1. Agriculture contributes to emissions by releasing substantial amounts of methane, nitrous oxide, and carbon dioxide. Consequently, agriculture may need to reduce emissions. We need to consider the cost from actions such as reduced fertilizer use, altered livestock feeding, and diminished rice acreage.

2. Agriculture provides a potential means for emission mitigation by offering opportunities for enhancing sinks.

3. Agriculture may be able to serve as an alternative source of fuel to displace fossil fuel combustion.

4. Agriculture may find itself operating under policies, designed to reduce global greenhouse gas emissions, that influence agricultural input and output prices.

5. Products from the forest sector compete in the market with other, often more energy-intensive, products like concrete, glass, and steel. Policies inspired by the Kyoto Protocol may change the demand for such products.

Policies that are agriculturally oriented and inspired by something like the Kyoto Protocol to reduce net greenhouse gas emissions need to consider the total effect of all of these roles that agriculture might play.

SUMMATION

Whether, and to what extent, carbon will be sequestered in forest and agricultural soils will depend on issues of practical economics and policy design. In some cases carbon mitigation will cause small alterations in current land uses, perhaps to the benefit of all. In other cases land for carbon mitigation may displace current land use practices. In yet other cases incentives may simply be insufficient to stimulate changes in land use practices. In general the establishment of a carbon trading market will provide additional economic opportunities for land-holders and rural communities. The big questions are:

1. Will agricultural soils be approved as means to meet greenhouse gas emissions commitments?

2. Will the incentives be there so that land owners will do this in their own best interests?

3. Can an emissions trading program be designed so the carbon that is paid for is kept for the needed future?

4. Can incentives be designed so that countervailing losses of carbon are not stimulated? In that vein, how does one avoid paying people who would never have done countervailing land transfers and how does one avoid a major private property rights issue?

5. How will emissions reduction be integrated into the total fabric of agricultural policy?

6. Are there agricultural or forestry practices that would increase net carbon storage in soils with positive, or small negative, impacts on productivity along with gains in externality accounts?

7. How will international agricultural activities come into play?

8. What is the relative ranking of soil carbon sequestration in terms of other possible agricultural actions that could be undertaken within the context of an emissions reduction program?

9. How will uncertainty in sequestration rates be taken into account?

Addressing these questions implies a multidimensional research agenda where one not only looks at the technological possibilities for carbon sequestration but also examines the design and structure of incentive and policy programs to stimulate sequestration activity. Such work must consider not only the direct effects of incentives and policies on the target activities but also the indirect effects across the economy that may lead to unexpected external costs, benefits, or sequestration offsets. Modeling will play a large role in addressing these questions, allowing researchers to examine both the technological and economic alternatives for sequestering carbon in terms of broad consequences for the agricultural system.

ACKNOWLEDGEMENTS

We acknowledge our debt to other people who have contributed to the ideas presented here. We particularly thank Darius Adams, Richard Adams, Ralph Alig, Mac Callaway, William Hohenstein, Bernhard Schlamadinger, and Stephen Winnett, along with many others, for contributing to our collective knowledge over the years. Norman Rosenberg, John Bennett, John W. McClelland, Debra Reed, and Ronald D. Sands contributed useful comments on this paper.

REFERENCES

Adams, R.M., C.C. Chang, B.A. McCarl, and J.M. Callaway. 1992. The role of agriculture in climate change: a preliminary evaluation of emission control strategies. In *Economics Issues in Global Climate Change: Agriculture, Forestry, and Natural Resources*, J.M. Rielly and M. Anderson, (eds). Westview Press: 273-287.

Adams, D.M., R.J. Alig, B.A. McCarl, J.M. Callaway, and S.M. Winnett. Forthcoming. Minimum cost strategies for sequestering carbon in forests. *Land Economics*.

Bolin, B., 1998, The Kyoto negotiations on climate change: a science perspective. *Science* 279:330-331.

Bureau of Oceans and International Environmental Scientific Affairs. Office of Global Change. 1997. *Climate Action Report. 1997 Submission of the United States of America Under the United Nations Framework Convention on Climate Change*. Department of State Publication 10496. http://www.state.gov/www/global/oes/97climate_report/. July.

Canada. 1998. In UNFCCC, Subsidiary Body for Scientific and Technological Advice. Report FCCC/SBSTA/1998/Misc.6/Add1, Additional submissions by Parties, at http://fccc.de.

Council of Economic Advisers, Executive Office of the President. 1998. *The Kyoto Protocol and the President's Policies to Address Climate Change: Administration Economic Analysis*. July.

Energy Information Administration, U.S. Department of Energy. 1998. *Impacts of the Kyoto Protocol on U.S. Energy Markets and Economic Activity*. SR/OIAF/98-03, October.

Energy Information Administration (EIA). 1998. *Annual Energy Review 1997*. http://www.eia.doe.gov/emeu/aer/. June.

IPCC/WMO/UNEP. 1996. *Climate Change 1995: Impacts, Adaptation, and Mitigation of Climate Change: Scientific-Technical Analyses*. Prepared by IPCC Working Group II. Cambridge University Press, Cambridge.

Isermann, K. 1994. Agriculture's share in the emission of trace gases affecting the climate and some cause-oriented proposals for sufficiently reducing this share. *Environ. Pollut.* 83:95-111.

Jerko, C. 1996. The economic potential of producing energy from agricultural biomass. MS Thesis. Texas A&M University, College Station.

Kyoto Protocol to the United Nations Framework Convention on Climate Change. 1998. Climate Change Secretariat (UNFCCC). http://www.cop3.de/. March.

Lal, R., J.M. Kimble, R.F. Follett, and C. V. Cole. 1998. *The Potential of U.S. Cropland to Sequester Carbon and Mitigate the Greenhouse Effect*. Sleeping Bear Press, Chelsea MI.

Manne, A., and R.G. Richels, 1998. The Kyoto Protocol: a cost-effective strategy for meeting environmental objectives? Unpublished manuscript, Stanford University, CA.

Marland, G., and B. Schlamadinger. 1997. Forests for carbon sequestration or fossil fuel substitution a sensitivity analysis. *Biomass and Bioenergy* 13:389-397.

McCarl, B.A. 1998. Carbon sequestration via tree planting on agricultural lands: an economic study of costs and policy design alternatives. Unpublished report, Texas A&M University, College Station.

McCarl, B.A., and U. Schneider. Forthcoming. Greenhouse gas emission mitigation: implications for agriculture. *Choices*.

McCarl, B.A., D.M. Adams, R.J. Alig. Forthcoming. Analysis of biomass fueled electrical powerplants: implications in the agricultural and forestry sectors. *Annals of Operations Research*.

McCarl, B.A., M. Gowen, and T. Yeats. 1997. *An Impact Assessment of Climate Change Mitigation Policies and Carbon Permit Prices on the U.S. Agricultural Sector*. Climate Change Policies and Programs Division, US Environmental Protection Agency, Washington, DC.

Metting, F. B., J.L. Smith, and J.S. Amthor. 1999. Science needs and new technology for soil carbon sequestration, this volume.

Mosier, A.R, W.J. Parton, D.W. Valentine, D.S. Ojima, D.S. Schimel, and O. Heinemeyer. 1997. *CH_4 and N_2O Fluxes in the Colorado Shortgrass Steppe: II. Long-Term Impact of Land Use Change*. ARS Report Number 0000071163. Agricultural Research Service, Washington, DC.

Post, W. M., R.C. Izaurralde, L.K. Mann, and N. Bliss. 1999. Monitoring and verifying soil organic carbon sequestration, this volume.

Reeburgh, W.S., S.C. Whalen, and M.J. Alpern. 1993. The role of methylotrophy in the global methane budget. In J.C. Murrell and D.P. Kelly (eds), *Microbial Growth on C1 Compounds*. Intercept, Andover, UK: 1-14.

Tyner, W., M. Abdallah, C. Bottum, O. Doering, B.A. McCarl, W.L. Miller, B. Liljedahl, R. Peart, C. Richey, S. Barber, and V. Lechtenberg. 1979. *The Potential of Producing Energy from Agriculture*. Purdue School of Agriculture. Final Report to Office of Technology Assessment, US Congress.

United Nations. 1992. *Framework Convention on Climate Change*. United Nations Environment Programme Information Unit for Conventions, http://www.unep.ch

Wigley, T.M.L. 1998. The Kyoto Protocol: CO_2, CH_4 and climate implications. *Geophys, Res. Letters* 25:2285-2288.

COMMENTARIES ON "SOIL CARBON: POLICY AND ECONOMICS"

The commentaries that follow have been edited slightly so as to highlight their key points. Although some or most of the commentators' suggestions were adopted by the white paper authors in their revisions, we have left them intact so the reader can gain an understanding of the discussions and dynamics of the process used at the workshop.

JOHN BENNETT, *Soil Saskatchewan Conservation Association*

No matter how accurate the science and monitoring of sequestration is, it will be farmers who play the most essential role in sequestering carbon.

Everyone at this conference is aware of the major contribution soils have to make in terms of CO_2 offsets. Without farmers this potential will not be realized.

In November of 1998 the Canadian Federal Minister of Agriculture stated that prairie soils could provide 8%–18% of the requirements needed to meet Canada's emission reduction targets.

What should be noted is that we have a 9 million acre carbon sink on the prairies that to date has cost the taxpayer virtually nothing. This was accomplished without incentives or subsidies.

The creation of this functioning CO_2 sink was not easy. If we examine how this was accomplished, we can learn how it came about and avoid costly mistakes.

In addressing the focus of this conference I hope I can put forward some ideas that may be useful in the context of enhancing the storage of carbon in soils from a farmer's perspective.

Let's explore three main themes: incentives and disincentives, education and extension (peer support), and research needs.

Incentives and Disincentives

Before we can approach this subject it is important to note that the pioneers or innovators of this "best management" method that sequesters carbon were motivated by a sense of stewardship. These were farmers to whom soil erosion by wind and water was unacceptable. Reducing or eliminating tillage helped to achieve their objective. This system in turn enhanced the organic levels of their soils. Greenhouse gas reduction was not even a remote consideration to these farmers.

Please recognize that removing tillage from a conventional farming system is not easy. It raises the required level of management skill considerably.

The moisture efficiencies and reduced fuel consumption offered potential profitability to early adapters who were prepared to accept the risks and management challenges that went along with this system.

We must address the management requirements and risks associated if we hope to realize the full potential of soils as a carbon sink.

In the province of Saskatchewan more than 20% of the land is currently under low disturbance or zero-till management. Given the present circumstances, my personal opinion is that any further gain will be slow. Very poor grain prices will most likely reverse this upward trend.

It should be understandable that at this moment most farmers are skeptical about carbon credit trading. Trading carbon operates on the same principle that gives Canadian farmers $1.20 U.S. a bushel in initial payment for their wheat and their American counterparts $2.20. Meanwhile both are paying $150,000 for combine harvesters.

Marland et al. correctly identify the 'carrot' challenges. Too small a carrot will create no interest and no commitment. Too large a carrot, interest but no commitment.

The ultimate carrot would be to develop best management zero-till farms that are more profitable than their tilled counterparts. This would create both interest and commitment. These best management farms already exist on the prairies but remember they also go hand in hand with a higher level of management and greater risk.

What are the disincentives in our structure that discourage the adoption of best management systems that sequester carbon? The shortcomings of crop insurance are a universal, legitimate concern on both sides of the border. These insurance plans are often based on "area average" yield. They seldom take into account the additional herbicide and fertilizer costs that increase yield potential in zero-till dryland farming.

Poor commodity prices and increased freight costs are also disincentives.

One provincial study of prairie freight rates concluded that a $13.00 U.S. a tonne increase would result in the increase of a million acres of summer fallow. These acres would be lost as a carbon sink.

Penalizing early adopters would also be a disincentive. The Kyoto protocol recognizes changes since 1990 but also sets the first accounting period of improvements between 2008-2012. One of our most pressing issues will be to see our contributions since 1990 get credit. If contributions since 1990 are not recognized the only way farmers like myself would see benefits would be to erase our gains with a massive tillage program and start over. From a personal perspective I would be more than a little annoyed if the costly mistakes we all made to sort out a functioning conservation system were penalized instead of rewarded.

Let's conclude this portion by saying that help in reducing the risk farmers face and making their farms more profitable would be the best possible incentive.

Education and Extension (Peer Support)

Education and extension assume a much more important role when the higher level of management required to zero-till is acknowledged.

In our conservation organizations, ACTS (Alberta Conservation Tillage Society), SSCA (Saskatchewan Soil Conservation Association), ManDak (Manitoba North Dakota Zero-till Association), and the SCC (Soil Conservation Council of Canada), we use the term "across the fence-line technology transfer."

It simply means that most extension and education happens between farmers. The conferences that these organizations put on blend the knowledge of farmers and researchers. These conferences are very well attended and encourage farmers to change the way they farm. They also provide a network of support as they begin to finesse a new way of farming.

Research Needs

There is no doubt that there is a real need for practical research. A close collaboration between leading farmers and researchers would go a long way in making best management systems that sequester carbon successful.

There are several applied research farms that are supported by farmers, industry and government. What is paramount is that the problems encountered in making these new farming practices be successfully addressed and that the information reach those that need it.

Summary

The best incentive to increase soil carbon sequestration will be a better farming system with reduced risks.

1) Education and extension will support the higher management levels needed in order to succeed.

2) Practical, applied research will address the challenges facing farmers as they take up new farming methods.

3) Remember, farmers must be involved in the policy development that affects them. After all, they are the segment of society that will manage the soil that sequesters the carbon.

JOHN W. MCCLELLAND, *National Corn Growers Association*

Marland, McCarl, and Schneider have provided an interesting and fairly complete overview of several of the economic and policy issues that will face agricultural decision makers and others involved in structuring and implementing a global agreement to mitigate greenhouse gas emissions. Developing policies and programs to mitigate emissions will not be simple from either a mechanical or political point of view. Despite the obstacles that lie ahead, it is important to begin developing a context and vocabulary that are common to the macro issue of climate change but understood by agricultural policy specialists. The disconnect between those active in the climate change debate and the large body of agricultural professionals, including farmers, has caused a great deal of mistrust and confusion and, I believe, is a major reason for holding this conference. Hopefully, the paper by Marland, McCarl, and Schneider is a beginning toward a more comprehensive understanding of the synergies between climate change and agricultural policies and programs.

Kyoto

The section concerning the Kyoto Protocol raises the most important point that will be discussed at this conference. It is important not to get hung up in the politics of treaty ratification but, rather, to look at Kyoto as a roadmap illustrating some of the routes that may be traveled. Kyoto is not *climate change*, and climate change is not *Kyoto*. They are not one-in-the-same and we must work to make sure thoughtful discussions about climate change are not overtaken by emotional arguments over provisions of the Kyoto agreement. This section of the paper also focuses on the parts of the agreement that are most directly relevant to agriculture as a medium for carbon sequestration. Two immediate questions come to mind here: what are the pros and cons for agriculture of being included in a tradable carbon credit program of whatever scope; and how to we measure the results of this program accurately and reliably?

Practices and Externalities

The next three sections of the paper speak to some fundamental questions in agricultural policy. First, is it better to develop policies that move toward greater productivity in existing land and also bring other land into productive utilization, or is it better to idle land? Second, what are the synergies with other policies and programs related to issues such as water quality, soil erosion, or agricultural research? How do we develop a system of coordinated research that maximizes the positive external economic effects and minimizes the negative?

These sections begin to lay out some substantial research questions that need to be addressed before effective policies can be developed. The authors give some indications of what these results might be, and I think they are important and encouraging. Encouraging because they point toward a more productive agriculture that uses science and technology to enhance the quality and quantity of output while reducing input costs. This is a track the corn industry is already on, and corn is particularly well positioned to meet the responsibilities of a low-carbon emitting world. The major life-science companies are already telling the corn industry about plans to deliver genetic material that can be engineered to meet product specifications without sacrificing yield, for example, a higher yielding corn with a stover that is more easily converted into ethanol that could be grown on many more acres that we use today because it is more drought and insect resistant.

Can We Make It Last?

Reducing carbon emissions through advances in mechanical and genetic technology and sequestering carbon in agricultural soils does have limits. Therefore any policy toward long-term carbon sequestration must be sustainable or it will not have the desired long-run effect. That is why there must be a two-pronged approach to carbon management in agriculture that maintains the gains made in the limited sequestration arena by linking them to carbon emission-reducing policies like using biofuels. Using biofuels always generates carbon emissions credits to the extent there is a reduction in net greenhouse gas emissions. Therefore farmers will be able to derive benefits by way of additional income for agricultural commodities they produce and are demanded by biofuels producers.

Creating that additional demand is the difficult policy issue facing biofuels. Some have suggested using carbon taxes to reduce the difference between fossil fuels and biofuels. Others prefer to use incentives to promote use of biofuels. I reject carbon taxes or most other schemes that impose taxes because I do not believe they will be enacted and any program that is going to rely on taxes for its success is doomed to failure. Tax incentives, on the other hand, can be argued as tax cuts that stimulate growth and create jobs. Yes, they create economic distortions, but we are already arguing about an imperfect market that has failed and created this huge externality of uncertainty about the global climate.

Whatever course is taken, we need to understand that for quite a few decades to come the world will continue to be faced with the scenario of cyclically low fossil fuel prices. Just as we are seeing a collapse in world oil markets today resulting from a decrease in worldwide demand, we could see low fossil fuel in the future as today's big oil consumers enact policies to reduce fossil fuel consumption. Thus, any policy to reduce fossil fuel use must have with it a long-term commitment that sustains it through those times when short-run profits conflict with long-run goals.

Verification

Verification appears to be a huge problem that must be solved before a worldwide carbon credit trading system can be introduced. It seems that the only logical way to develop a verification system that can do a reasonable job of estimating the benefits of carbon sequestration from a given set of practices or the implementation of a given set of programs is to build a model-based system. This type of certification is done in some programs now like the federal reformulated gasoline program. In that program, the performance of batches of refinery gasoline is certified using a statistical model. The model does not tell us how a particular set of hydrocarbons will react in a specific automobile to produce emissions, but how general fuel characteristics in a reformulated gasoline reduce emissions from a representative fleet of vehicles when compared to a baseline gasoline. It seems plausible that enough scientific information about the effects of agricultural practices on carbon sequestration to develop a modeling system that would provide verification of carbon reductions exists or could exist with a well focused research effort. Clearly, the massive amount of data USDA collects from farmers about their practices and the monitoring of program compliance would make it possible to produce such a model for the U.S. Globalizing such a model or at least extending it to Annex I countries could be a major contribution of U.S. agricultural science.

Income and Economics

One reason agriculture should take a serious look at climate change issues is because of the potential impact adoption of different policies could have on farm income. This includes the possibility of significant negative effects on farm income if carbon taxes increased fuel and fertilizer prices by 30-50 percent in a relatively short period of time as some studies have suggested. But the agricultural community should also consider the positive effects on farm income that are possible under other scenarios. As the authors point out, agriculture has fought a losing battle against an inelastic demand curve for traditional agricultural products. Some climate change policies could create new opportunities for production agriculture that would allow farmers to start producing crops that have a more elastic demand curve, or make the demand curve for traditional agricultural products more elastic. An example of the latter is using corn to produce ethanol.

Determining how to proceed is largely a function of research. What can we produce and process at the least cost? The answer is probably a broad mix of crops that are used for both food and fuel. However, we must first settle the issue about what our long-term policy regarding climate change is going to be, because this policy will determine where potential economic returns will be available for farmers and others in production agriculture and processing. Making informed policy choices for agriculture and other segments of society requires knowledge of complex information and the ability to communicate this information effectively to the public. Farmers are willing to discuss this future and be part of the solution, but farmers are not willing to pay the costs of reducing greenhouse gas emissions without obtaining any of the benefits.

DEBRA REED, *Office of Senator J. Robert Kerrey*

The authors provide broad coverage of many of the complex economic and policy issues surrounding the topic, offering a vehicle for productive debate and sharing of perspectives. I have limited my remarks to three broad, overarching areas, with additional points on two specific issues. The first area is what I would call the Kyoto opportunity, or the opportunity presented to agriculture to participate in the solution of the global climate change problem. In addition, I have addressed some aspects of the associated international and domestic issues involved. Finally, I offer observations about the issues of externalities and retention of stored carbon.

The Kyoto Opportunity: Agriculture as a Bridge

In the context of the Kyoto Protocol (or a Kyoto-like treaty), the issue of enhancing carbon storage in soils presents opportunities to the U.S. agricultural sector and to the broader international community. As advocated by Pacific Northwest National Laboratory (PNNL), soil carbon sequestration can provide the bridge to help achieve emissions reduction commitments at the lowest possible cost to other sectors. The underlying political (and economic) issues, however complex, are not insurmountable.

The recent decision taken at COP4 in Buenos Aires related to land use, land use change and forestry, represents progress towards achieving the reality of credits for soil sinks. The eventual outcome of this decision should be adoption, at COP6 in 2000, of the definitions, rules and mechanisms whereby soil sequestration activities will "count" under the Protocol. Ensuring the completeness and robustness of the data in the IPCC Special Report on Articles 3.3 and 3.4 is essential to this outcome.

The critical policy issue now is development of a cohesive framework for the agricultural sector as a whole to take advantage of the opportunities presented by the Protocol. Short-term, and in the two years leading to an anticipated decision on sinks at COP6, agriculture can position itself as "the bridge," to ensure acquisition of credits and incentives. It will take the combined efforts of the sector, particularly in the political arena, to achieve full potential domestically and internationally (although domestic credits can precede an international system, and linkage of the two is not necessarily absolute—that is, domestic credits in the absence of international credits are possible). Particularly if a U.S. credit for early action program is ever enacted, public-private cooperation in the agriculture sector will be crucial; other U.S. sectors (buildings, industry, and transportation) are positioning themselves now for this possibility.

International Issues

Internationally, it will be necessary to inform other countries of the possibilities for soil carbon sequestration to pave the way for adoption of the language necessary for inclusion of soil sink activities under the Protocol. Preparation of the IPCC Special Report on Articles 3.3 and 3.4 of the Protocol, and the associated workshops, will aid in that endeavor. Desertification and land degradation provides considerable opportunity for interest and involvement by developing countries, particularly in Asia and Africa, the continents most affected and thus most likely to benefit from a reversal of these circumstances and the accompanying impacts. Inclusion of soil sinks in a treaty, however, will require comprehensive baseline accounting of sinks in each participating country and accepted methods of measurement, verification, and monitoring. These should be primary goals of the agricultural sector in the United States.

Domestic Issues

U.S. agriculture produces the safest, most abundant, most economical food supply in the world, and U.S. farmers are stewards of the land and of natural resources. Yet there is an increasing disconnect with the public, and even between scientists and policymakers, of the roles that farmers play in quality of life and environmental issues. The direct and ancillary benefits of soil carbon sequestration—to agricultural productivity and sustainability, air and water quality, reduced soil erosion, biodiversity and habitat—are societal benefits that should be quantified and rewarded. Regardless of what is negotiated under future international agreements, incentives and policies to reward agriculture can be crafted now, in any combination of three forms: tax incentives, program benefits, and, eventually, monetary payment for credits (for example, in a carbon trading scheme).

The authors also broach farm income supports and question whether the Protocol opportunity provides justification for a new breed of farm programs. The answer: absolutely. Carbon sequestration can provide the nexus for such a paradigm shift. The ancillary benefits alone provide rationale for incentives and credits; carbon sequestration is an added bonus. Education and appropriate characterization of the benefits (in the political arena) can help to achieve the adoption of supportive policies and programs.

To maximize potential domestic (and eventually, international) credits for agriculture, a productive policy path would be a transitional phase focusing on sustainable land use agricultural policies and programs. The addition of a carbon element or factor to existing farm programs, where the benefit exists (CRP, EQIP, WRP, etc.), and incentives for management practices such as conservation tillage can be reinforced as we learn more. This path would allow program and tax incentives to be granted initially, with later potential supplementation or replacement by free market trading.

The knowledge and information exists now to begin adapting agricultural programs to provide incentives, and through careful attention to construction of programs it is feasible to create tradeable carbon credits if/when a domestic free-market system becomes operational. A limited number of full-scale pilot projects to test the trading concept under various farm scenarios would bolster longer-term efforts and help with the learning curve. Project-based efforts are already underway in the U.S. and in Canada. Uncertainties in soil carbon measurements can be dealt with at the market level and should not detract from or delay inclusion of soil projects in trades.

Externalities: Long-term planning can avoid many negative externalities. For instance, the authors reference potential deleterious effects on the traditional forest sector caused by programs designed to move agricultural lands into forestry; the case underscores the dangers of not developing an agricultural framework and of perhaps adopting or advocating land use change decisions before we know what will count internationally and before we can adequately model the effects of proposed changes. Similarly, the notion that a focus on carbon sequestration might lead to decreased food and fiber production is plausible only if we allow short-term or short-sighted decisions to drive agricultural policies. Faced with a burgeoning global population and projected regional food and water shortages, long-term agricultural policies and incentives should balance competing and complementary objectives alike. Additionally, soil carbon sequestration, at least from a food production viewpoint, can be complementary, not competitive. Lastly, in the globalized economy in which we operate, export opportunities should always remain a focus of U.S. agricultural policies.

Retention of Stored Carbon: The authors' theme title should be augmented to: "Longevity of agricultural carbon and keeping carbon you have *been* paid for." Retention should be the responsibility of the credit recipient, and retention periods should be factored into the payment contract. Thinking in Protocol terms, it is probably realistic to factor retention periods in terms of budget periods or 5-year time frames. Farmers or others who are paid to store carbon should expect to have to retain stored carbon for one or more budget periods, or else be debited for losses. And just as one can own a sequestration credit or a sequestration right, one should be able to resell the right. A precedent pointed out by Richard Sandor of the Chicago Board of Trade is that land is frequently sold decoupled from its mineral rights.

The two final points—about externalities and retention of carbon—are more logistical items that, like others, can be worked out with practice and implementation. The bigger issue now is the need for political will within the agricultural sector to move down this road, and to seize upon the Kyoto opportunity. Done correctly, carbon sequestration can provide the opportunity to weave a more stable safety net for agriculture, one that reaps multiple environmental rewards for society at large.

RONALD D. SANDS, *Pacific Northwest National Laboratory*

The authors of Paper 4 provide an introduction to 14 themes on the policy and economic dimensions of carbon sequestration in agricultural soils. These themes cover a very broad range of topics and are intended to stimulate further discussion. The paper provides a fairly thorough discussion of economic issues within the agricultural sector, especially with regard to designing an incentive system to promote carbon storage and the related problem of carbon leakage. The paper also does an especially good job of describing the issue of carbon longevity in Theme 5.

I will discuss the role of economic models and expand on Theme 8, Energy Policy and Carbon Sequestration, and Theme 10, Practical Economics. The text in both of these themes points out that we cannot deal with carbon sequestration in soils as an isolated topic. Theme 8 reminds us that we cannot separate energy policy from a policy to sequester carbon in soils. Theme 10 points out that carbon mitigation opportunities in the agricultural sector would compete with opportunities in the energy sector.

I will first comment on the role that economic modeling can play to inform policy decisions. Second, I will provide an example from recent work presented at the August 1998 meeting of the Energy Modeling Forum. I will try to make the case that top-down economic models cannot adequately handle carbon sinks without information from specialized studies in the agricultural sector. I will also argue that agricultural specialists should care about results from top-down economic models.

Top-Down Economic Models

Generally, economic models are constructed to answer particular questions, and a model designed to answer questions related to the Kyoto Protocol or other sets of emissions constraints should do at least the following:

1. Project baseline carbon emissions over time for a country or group of countries.
2. Find the least-cost way to meet any particular emissions constraint.
3. Provide a measure of the carbon price, in dollars per metric ton.
4. Provide some measure of the overall cost of meeting an emissions target.

The second and third goals are relevant to our discussions of sequestering carbon in soils. It is not possible to determine a least-cost strategy without understanding the opportunities available in all sectors, including opportunities provided by carbon sinks. However, it is likely that the carbon price resulting from any carbon permit trading scheme will be driven primarily by activity in the energy sector. Activities to sequester carbon in soils at or below the carbon price should be considered.

Energy Modeling Forum Scenarios and the Price of Carbon

What is a reasonable range for the carbon price? This was one of the questions posed to economic modelers participating in a recent modeling exercise to estimate the costs of complying with the Kyoto emissions targets. This exercise is coordinated by Stanford University's Energy Modeling Forum (EMF).

As pointed out in Theme 10, the cost of carbon emissions reductions depends on the extent of permit trading between countries. The following table shows just how wide this range can be, with almost an order of magnitude difference in the carbon price between the no-trading case ($188 per metric ton) and full global trading ($22 per metric ton). These results were obtained from the Second Generation Model (SGM), an economic model developed at Pacific Northwest National Laboratory. Of course, these numbers vary across models participating in the Energy Modeling Forum study, but the pattern is the same. One of the strengths of the EMF process is the inclusion of many models.

Round #2 Post-Kyoto EMF Scenarios
Second Generation Model

US Permit Prices in 2010
(1990 US$)

	No Trading	Annex I Limit on Sales	Annex I Limit on Purchases	Full Annex I Trading	CDM	Double Bubble	Annex I plus China and India	Global Trading
Kyoto Targets	**188**	181	103	**85**	73	59	27	**22**

It should be pointed out that these results do not reflect options available from reducing other greenhouse gases or from carbon sinks. When the EMF study was initially proposed, it became obvious that we could not proceed without specialized information on the costs of these options.

The Administration Economic Analysis went one step further by including supply curves for the other greenhouse gases, using information provided by the U.S. Environmental Protection Agency. However, carbon sinks were not included in the Administration analysis.

Supply Curves for Carbon Sinks

It is not enough to simply list the various sequestration options and their technical potential. We need a way to prioritize options based on cost. A supply curve is then constructed by starting with the least expensive options and then including the more expensive options in sequence. This provides a way for the economic models, and for policy makers, to determine which sequestration options should be considered, for any given carbon price. It also provides a way to compare costs of sequestering carbon in soils with options from the other greenhouse gases and the energy sector. Again, supply curves for sequestering carbon in soils can only be generated using information provided by specialists in the agricultural sector.

Summary

Economists who build and operate top-down models are certainly interested in including carbon sequestration options in their analysis. This requires a dialogue with other modelers specialized in the economics of agriculture and in the processes governing carbon storage in soils. Economists are interested in finding a least-cost strategy of meeting emissions constraints, which should include all practical options. Agricultural specialists should care about how agriculture is represented in top-down analysis that estimates overall costs and ultimately informs policy.

Working Group 1: New Science

Submitted by John Reilly, *Chairman*

To meet our charge we first addressed these questions:

How good is the current science and technology of soil carbon sequestration—what do we know now? What are the promising potential new technologies? What are the science problem areas? How do we categorize and organize science problem areas and research? How can we determine priorities among identified research areas? What is the probability that any one of these will pay off? What makes a successful technology? And perhaps most critical—how do we assess the overall potential of soil carbon sequestration in comparison with other climate change mitigation approaches which also require new science and which will, necessarily, compete for the same research funds?

We agreed in principle that the working group was not fully equipped to answer all of these questions, but rather that their full assessment was, in itself, worthy of research and immediate attention. Nonetheless, a hierarchy and priority of science questions was formulated.

1. What is the geographical distribution of land use across predominant terrestrial ecosystems and in which of these resides the potential for land use change toward ecosystem types that store more carbon?

2. What are the factors that control Net Primary Productivity (NPP) and Net Ecosystem Productivity—the latter determining the amount of biomass that can potentially be retained in soils and standing biomass?

3. What are the factors that control decomposition of organic matter and release of carbon from soils? How can the residence time of carbon in soil be lengthened?

What follows is a set of recommended research priorities that address not only these three science questions but also some economic questions, the answers to which will determine whether soil carbon sequestration programs are actually implemented

RECOMMENDATION 1

Identify the factors that affect residence time of carbon in the soil and the ways in which soil carbon quantity and residence times are affected by ecosystem dynamics.

Rationale: There is much we do not know about the basic processes that determine "equilibrium" or pre-disturbance levels of carbon and its long-term storage potential in soils. The dynamics of aggregate formation and the factors that determine aggregate persistence in soil are also not fully understood. An immediate focus on soil ecology, microbial ecology, the role of plant communities, and interactions between nitrogen and carbon cycles is needed.

Much is known about ecosystem dynamics with regard to soil carbon, but this knowledge has not been organized effectively to focus research on increasing the residence time of soil carbon. Science of the rhizosphere and morphology of roots needs more attention. We must also increase our understanding about the size of current carbon pools and how carbon is reallocated among labile and non-labile pools.

RECOMMENDATION 2

Systematize knowledge of land restoration processes—including economic consideration

Such knowledge will lead to a better understanding of the amounts of soil carbon sequestration possible, quantities of wood and grass that can be produced, most effective practices, nutrient retention, effects on water quality, and other environmental consequences.

Rationale: The main potential of carbon sequestration as understood today is to restore carbon stocks in soils of degraded lands to near their pre-disturbance levels. Practices that can restore productivity (afforestation, reforestation, reseeding grasses, etc.) can also have environmental consequences—probably, but not always, benign. These must be factored into land restoration activities whether or not soil carbon sequestration is the primary purpose of these activities.

RECOMMENDATION 3

Improve the scientific basis for measurement and verification

Advances must be made in in-situ, non-destructive and inexpensive soil carbon measurement methods able to detect small changes over time.

Rationale: There is need in the near future to develop a system of accounting for emissions and capture of carbon that will be acceptable to the Parties to the Kyoto Protocol. Improved monitoring methods could reduce the cost of this accounting, limit conflicts over what is actually being sequestered, and reduce the potential for "leakage."

RECOMMENDATION 4
Understand economics of soil carbon at the enterprise level

Rationale: Before farmers and forest managers will be willing to engage in carbon sequestration activities, they will need to understand the financial consequences of adopting specific practices. If an option is not profitable under current economic conditions, the incremental costs associated with it must be known for farmers to establish the price they can charge for carbon storage or the incentives they will need to engage in a carbon storing activity. In their calculations, farmers will need to consider these and other effects of carbon sequestration activities on farm profitability and how these compare with the best alternative farming systems. Other, non-market benefits and costs such as nitrate leaching and water quality, chemical residues, erosion and water body sedimentation must also be considered.

RECOMMENDATION 5
Develop supply curves

The supply curve, a tool used by economists, can be applied to provide answers to policy questions as well as guidance for setting research priorities. To what extent can soil C sequestration serve as a bridge to the future? How much additional soil C sequestration can we really put in place by 2008–2012? An improved understanding is also needed of the dynamic path of carbon sequestration, given basic limits in NPP and ultimate limits on soil carbon sequestration.

Rationale: Research funders and policy makers need to have firmer estimates of the economic potential for soil carbon sequestration. However, care is needed to relate the supply curves to the dynamics of new technologies, the sequencing of their introduction, and their interactions. In developing the longer-term supply curves, analysts must consider the possible feedback effects of changing climate as well as potential changes in the agro-economic-land use system that might occur unrelated to climate change.

RECOMMENDATION 6
Use consistent scaling methods

Scientific data and interpretation is needed to establish the 1990 baseline values of carbon in the soils of the United States and other nations. For baseline values to be credible, results from a wide range of research studies and other observations must be scaled to broader areas in a consistent manner.

Rationale: Lack of consistency in approach contributes to a large volume of noise in the literature. It is unclear how or whether studies at the individual plant, microbe, or field level scale to larger areas. What are the controlling factors that act at broader scales?

RECOMMENDATION 7
Take a systems perspective

A systems perspective is needed for short-term "alternative management systems" as well as more aggregate and longer term analyses to consider ancillary benefits and costs.

Rationale: Some of the complex interactions in soil carbon dynamics would lead to improvements in other dimensions of agro-forestry-environment interactions and some could have negative effects. We need to look for the unexpected.

RECOMMENDATION 8
Develop engineering technology

We need to develop the engineering technology that would allow implementation of options including, for example, water delivery, energy efficiency of farming systems (and input production, delivery), and nutrient technology. Since much of this is applied product development research, we need also to consider what would stimulate and encourage the private sector to undertake this research.

Rationale: Carbon sequestration in soils will not occur unless the equipment and supplies are available at a cost that makes the systems profitable for farmers. Low energy efficiency can "discount" the value of carbon sequestration by offsetting energy emissions from the farm or upstream manufacturing. Political reality is that Congress is not likely to fund research for applied/commercialization research. Commercial reality is that industrial firms have the knowledge of what they can sell to farmers and how to sell it.

Working Group 2: Monitoring and Verification

Submitted by Julie Jastrow, *Chairman*

Those responsible for developing policy and international agreements to mitigate climatic change must be convinced that potential changes in soil carbon may be large enough to be worth including as carbon sinks in future agreements. Because decisions will be made within the next 18-24 months, this is an urgent task. To accomplish it, we must demonstrate that (a) it is possible to measure soil carbon with sufficient accuracy and precision to detect changes and (b) the methods to sample, monitor, and verify these changes are cost effective. Such demonstrations will be necessary to help secure the acceptance of soil carbon sinks by other nations. There are several existing projects that can be used as examples, including the Canadian project initiated in southern Saskatchewan by GEMCO[1] and the work of Colorado State University and its collaborators on long-term agricultural sites, modeling, and spatially explicit regional accounting. In addition, there are many examples in the literature of documented changes in soil carbon resulting from changes in land use or management practices.

Once the potential significance of soil carbon sinks is recognized by policy makers, monitoring, verification, and predictions of changes in soil carbon storage at various spatial scales from the individual field to national levels will be necessary in order to implement projects or obtain credit for changes in national inventories. We recommend the following actions to facilitate development of monitoring and verification strategies and protocols.

[1]GEMCo (Greenhouse Gas Emissions Consortium) is a Canadian consortium of energy and chemical manufacturing companies committed to act on the issue of climate change. Consortium members have implemented their own corporate greenhouse gas action strategy and together have identified the need for an integrated and cooperative approach in research, planning and project implementation to effectively address global climate change. Members are actively participating in domestic and international greenhouse gas offsets. GEMCo is currently co-funding a field project in Canada to verify carbon sequestration in agricultural soils.

RECOMMENDATION 1

Use and improve current methods/programs to monitor and verify changes in soil carbon

Existing information, databases, sampling/analysis methods, and statistical procedures for obtaining adequate samples can be used for developing initial carbon sequestration projects and programs to monitor and verify changes in soil carbon. Efforts to improve these capabilities, reduce the errors inherent in measurements of soil carbon stocks and fluxes and, thereby, improve our ability to detect changes are ongoing and should continue with added support. As more knowledge is gained and better methods are developed, procedures and methodologies can be adapted and improved.

RECOMMENDATION 2

Focus initial sequestration projects on systems with the highest expected gains in soil carbon in order to increase the probability of detecting and verifying these gains

Currently, afforestation, the application of best management practices in agricultural systems (e.g., sustainable rotations, no-till), and the conversions of annual systems to perennial systems (e.g., the Conservation Reserve Program (CRP) and biomass cropping) probably hold the most promise. However, an initial focus on these systems should not limit any agreements regarding soil carbon sinks to specific management practices and/or land-use changes. Rather, care should be taken to ensure that any agreements with regard to carbon sequestration are general enough to allow the future inclusion of currently unknown or apparently unimportant sinks. Development of new measurement capabilities and/or management practices may make such sinks observable and consequential at a later time.

RECOMMENDATION 3

Establish representative benchmark monitoring sites, making use of traditional field sampling procedures and/or eddy flux measurements

Such sites are needed to provide documentation of actual changes in soil carbon content for research purposes, specific projects, and field validation of model predictions. Information from a network of project-based sites could also contribute significantly to national inventories.

RECOMMENDATION 4

Use models to scale from the field to larger areas, including the nation

It is patently impossible to measure changes in soil carbon content in every field of every project or in every county or other political or natural division of a nation. Therefore, a range of models—including empirical, process, and meta (e.g., look-up tables or expert systems derived from empirical or process models)—will be required at both the project and national levels to predict changes in soil carbon stocks. Scaling up to project/regional/national levels (stratified accounting) will require applying modeling results to spatial areas at some level of aggregation appropriate to the project or inventory at hand, based on factors such as land use, topography, soil type, or watershed. Measures of temporal variability in parameters such as primary production (carbon inputs to the soil), decomposition, and erosion should be applied to spatially explicit model predictions of soil carbon dynamics to better simulate, predict, and adjust for temporal responses to variations in external forcing factors (e.g., climate, fire, storm damage, disease, pest damage). Although temporal responses in organic inputs may be modeled from variations in external forcing factors, they may be best accounted for by using available information on plant growth, such as yield or remotely sensed data on net primary production (NPP). Similarly, variations in external forcing factors, especially climate, may be used to adjust decomposition rates, and variations in erosion rates can be accounted for by current predictive techniques (e.g., the Revised Universal Soil Loss Equation).[2]

Nevertheless, our ability to accurately scale up to larger spatial areas needs to be improved. Outside the United States, Canada, and parts of Europe, a major problem constraining stratified accounting approaches may be the lack of appropriate spatially referenced databases. Where such databases exist, work is needed to determine how to best define aggregated area units in order to control scaling errors. If area aggregates are additive, then errors associated with over- or underestimation should average out. However, interactions across or between area aggregates (e.g., erosional losses in one aggregate that are deposited in another) could lead to potentially significant errors. Research is needed to assess the extent of such interactions and to develop methods to either account for them or eliminate them.

On a smaller scale, meta models that can run on personal computers need to be developed to permit predictions of soil carbon storage for small projects, in particular. Such models might be derived from empirical relationships or could be expert systems that utilize response surfaces derived from the output of process models (i.e., essentially, look-up tables). In either case, the model would operate by inputting (a) relatively easy to obtain or measure environmental and biological/production data and (b) answers to a series of relatively simple questions (e.g., to identify management practices).

RECOMMENDATION 5

Improve the science and technology of monitoring and verifying changes in soil carbon

Although we currently have acceptable methods to measure changes in soil carbon and reasonably good models of soil carbon dynamics, we need to continue efforts to improve our basic scientific understanding of the mechanisms of soil carbon sequestration and how these mechanisms control carbon cycling and turnover. Such basic research will help to improve the parameterization of existing models and may even lead to the development of new models constructed around measurable soil carbon pools rather than the conceptual pools currently employed by most soil organic matter models. Similarly, work is also needed to improve the

---

[2]Revised Soil Loss Equation: a computerized program to calculate soil loss based on rainfall-runoff erosivity, soil erodibility, slope length and steepness, management, and conservation practice.

applicability, fit and predictive capabilities of soil organic matter models for a wider variety of regions, ecosystem types, and management practices.

Increased efforts to couple eddy flux measurements to soil carbon storage are needed. Ecosystem carbon fluxes and stocks (including soil carbon) need to be measured at eddy flux sites to determine if eddy flux measurements of net ecosystem production can be used to predict changes in soil carbon storage.

Research is needed to determine how management changes across various land use types and climatic regions affect the production of other greenhouse gases (e.g., N_2O, CH_4) to ensure that emissions of these gases do not negate the benefits gained by the uptake and sequestration of CO_2. Similarly, the carbon costs (e.g., energy use associated with fertilizer production and application) of management changes need to be assessed from an ecosystem perspective. Although it would be impossible to explicitly measure other greenhouse gas emissions and account for other carbon costs for every project, research is needed to evaluate the potential effects of specific practices in specific regions and to identify practices for which the costs are too high.

Development of an accurate, portable, non-destructive organic carbon analyzer that can be used relatively quickly by one person in the field would be extremely helpful for monitoring and verification activities. The analyzer cannot simply quantify surface reflectance but would need to be capable of probing by depth increments and integrating within some known volume. Ideally, the analyzer would also be capable of measuring nitrogen. Such an analyzer would allow repeated sampling within relatively small areas without the removal of soil from the field, thus leaving the study area intact. In addition, to prevent the significant compounding of errors, very accurate estimates of soil bulk density will be required for converting measured carbon and nitrogen concentrations to standing stocks (masses per unit volume or area).

RECOMMENDATION 6
Design an effective monitoring system

An effective and economical monitoring system will be needed to assure compliance with national and enterprise-scale commitments to soil carbon sequestration practices. Independent, random checks will be required to assure the credibility of the reporting of sequestration activities and the measurement of changes in carbon stocks. A variety of tools, including remote sensing, may be appropriate for such activities, and the program should be capable of adapting to take advantage of improvements in monitoring and measurement technologies.

The selection of appropriate methods and development of acceptable protocols for monitoring and verification procedures will require much effort, but can be done. Creation of specifically focused working groups and workshops will be required to evaluate potential methodologies and recommend procedures. Ultimately, some sort of national panel with the requisite expertise (e.g., sampling, statistics, measurement, modeling, GIS, remote sensing) may be needed to select methods and protocols and review proposals for new methods and approaches.

Working Group 3: Desertification

Submitted by William Pennell, *Chairman*

There is great potential for removing carbon from the atmosphere by sequestering it in degraded range, farm, and forestlands in the form of soil organic carbon (SOC). According to Lal et al.,[1] about 17% of the Earth's surface has been degraded by human activity. As a result of this degradation, soils have lost significant quantities of SOC—on the order of 20-30 Pg.[2] Restoring all or a major fraction of this SOC would sequester a quantity of carbon equal to about ten years of the atmosphere's current rate of increase.[3] In addition to offsetting a significant fraction of the current anthropogenic emissions of carbon, restoring SOC in degraded soils would have a substantial positive impact on the quality of life of peoples living on these lands by improving soil productivity and reducing erosion.

The ecological and human imperatives for reversing desertification and improving the condition of degraded lands have been recognized for decades, and they are quite independent of the role these practices could play in combating climate change. In spite of this knowledge, the fraction of the Earth's surface subject to desertification continues to increase. The problem is not technical. There are well-known and effective land management practices that can improve soil condition, combat erosion, and increase SOC. The problem is social—economic, political, and cultural. It is obtaining the money required to implement land restoration on the scale required, and doing this in countries whose civil and political institutions are weak or chaotic, and engendering a land stewardship ethic in those who actually use the soil.

[1]See Lal, R., H.M. Hassan, and J. Dumanski, "Desertification Control to Sequester C and Mitigate the Greenhouse Effect," this volume.

[2]Ibid.

[3]Cole, V., C. Cerri, K. Minami, A. Mosier, N. Rosenberg and D. Sauerbeck. 1996. Agricultural options for mitigation of greenhouse gas emissions. Chapter 23 in R. T. Watson, M. C. Zinowera and R. H. Moss (eds), *Climate Change 1995. Impacts, Adaptations and Mitigation of Climate Change*. Report of IPCC Working Group II, Cambridge University Press: 745-771.

How does soil carbon sequestration change this picture? Why do we think that a worldwide program of improvement in the condition of degraded lands can be implemented now, when efforts at reversing desertification have been largely unsuccessful in the past? First, making a commodity of carbon credits could greatly increase the amounts of money available for land reclamation efforts. Second, developed countries would have an interest in seeing that land restoration projects are initiated and sustained so that they can obtain carbon emission offsets. In other words, something could happen if a mechanism is established through which carbon sequestration generates a conditional flow of money to those who simultaneously stabilize and improve degraded land while sequestering carbon and mitigating the greenhouse effect.

In light of these assumptions, the working group made the following recommendations.

RECOMMENDATION 1

Develop a mechanism to establish a market for tradable carbon emission credits to be applied to desertification control and restoration of degraded lands

As indicated above, implementation of an extensive program of increasing SOC in degraded lands can lead to the sequestration of large quantities of atmospheric carbon, offsetting a substantial fraction of current anthropogenic carbon emissions. Economic modeling studies have shown that a free market in tradable carbon emission credits could be a very effective method for motivating activities that reduce net greenhouse gas emissions and minimize their costs.[4]

RECOMMENDATION 2

Link activities that provide carbon sinks with the clean development mechanism

As part of the Kyoto agreement, a "clean development mechanism" was established by which interests in developed countries can obtain carbon emission offset credits for implementing projects in less developed countries that reduce carbon emissions through the use of modern technology. This program could be attractive to developing countries, giving them access to advanced technology. Linking soil carbon sequestration programs, which in many cases do not require advanced technology to implement, to the clean development mechanism could provide additional mechanisms for initiating and funding them. In order to sustain this option it is very important that the clean development mechanism be institutionalized so

[4]Edmonds, J., M. Wise, H. Pitcher, R. Richels, T. Wigley, and C. MacCracken. 1996a. An integrated assessment of climate change and the accelerated introduction of advanced energy technologies: an application of MiniCAM 1.0. *Mitigation and Adaptation Strategies for Global Change* 1(4):311-339; and Edmonds, J., M. Wise, R. Sands, R. Brown, and H. Kheshgi. 1996b. *Agriculture, Land-Use, and Commercial Biomass Energy: A Preliminary Integrated Analysis of the Potential Role of Biomass Energy for Reducing Future Greenhouse Related Emissions*. PNNL-11155. Pacific Northwest National Laboratory, Washington, DC.

that it can be developed and implemented, no matter what the fate of the current Kyoto agreement.

RECOMMENDATION 3

Establish an international carbon fund

It may not be possible to fund programs of a size that could sequester enough carbon to have global impact through private investment projects alone. This could be especially true in countries lacking well-developed market-based economic systems or legal systems for protecting and enforcing contracts and property rights.

RECOMMENDATION 4

Restore degraded marginal lands through afforestation and revegetation and intensify management of vegetation on prime farm and rangelands

Land use practices in many parts of the world are mining soil carbon for the production of food, fuel, and fiber. As we improve the conditions of marginal lands, we must also intensify agriculture on the most productive lands. Increasing agricultural productivity can be accomplished in ways that are sustainable and improve the condition of the soil. Improved productivity is essential for meeting human needs and for preventing desperate populations from raiding marginally productive lands for subsistence.

RECOMMENDATION 5

Establish large-scale pilot projects for demonstration and benchmarking in different countries and representative ecosystems

Pilot projects are needed for on-site demonstration of and training in land use practices for improving soil condition and productivity. These projects should cover a variety of climatic and soil types in regions of diverse economic and social conditions. These projects can also serve as sites for observation of soil carbon retention as affected by climate, soil and management practices.

RECOMMENDATION 6

Develop methods and infrastructure for determining country-by-country baseline data on soil carbon stocks and fluxes

Improved information on existing soil carbon stocks and fluxes will be essential for implementing a large-scale carbon sequestration program.

RECOMMENDATION 7

Develop in-country capacity for implementation of emission reduction and carbon sequestration offset projects and for measurement and monitoring of relevant carbon pools (soils and vegetation) and fluxes

In order to be effective over the time scale required for carbon sequestration, these projects must become internally institutionalized.

RECOMMENDATION 8

Incorporate local expertise regarding land conditions and management

In many situations traditional land-use practices have evolved that maintain long-term soil productivity. This local knowledge must be the basis of any program for land restoration.

RECOMMENDATION 9

Adopt a participatory approach in implementing projects and, where appropriate, use farmer-to-farmer technology transfer

The people who are actually using the land must feel ownership of the new practices that are being advocated. Training and advice is most effective when it comes from those who look like, act like, and talk like you do.

RECOMMENDATION 10

Develop mechanisms for vesting long-term interest in the condition of the land resource

Inequitable land tenure arrangements will impede efforts to implement continuity of soil improvement and land restoration projects. People will not invest labor and capital in improvements on land for which they have no responsibility or that they may lose before they can reap the benefits.

Appendix I

Participants in the April 23-24, 1998, Planning Meeting

Pacific Northwest National Laboratory
Harvey Bolton
John Clarke
Jae Edmonds
Elizabeth Malone
Blaine Metting
Richard Moss
Cesar Izaurralde
William Pennell
Steve Rayner
Norman Rosenberg

Oak Ridge National Laboratory
Jeffrey Amthor
Michael Farrell
Steve Hildebrand
Gary Jacobs
Russell Lee
Wilfred Post
Sandy McLaughlin
Lynn Wright

Brookhaven National Laboratory
George Hendrey

US Department of Energy
Roger Dahlman
Jerry Elwood
Jeff Summers

US Department of Agriculture
Herman Mayeux
Tim Parkin
William Sommers
Elvia Niebla

Universities
Lee Burris, *Iowa State*
Don Duvick
Rattan Lal, *Ohio State*
Bruce McCarl, *Texas A & M*
Keith Paustian, *Colorado State*

Others
Gunter Reithmacher, *World Bank*
Richard Stuckey, *Council on Agricultural Science and Technology*

Appendix II
Agenda

Carbon Sequestration in Soils: Science, Monitoring and Beyond

December 3-5, 1998 Harbourtowne Resort St. Michaels, MD

A workshop organized by

Pacific Northwest National Laboratory (PNNL)

Oak Ridge National Laboratory (ORNL)

Council for Agricultural Science and Technology (CAST)

with the support of

U.S. Environmental Protection Agency (USEPA)

National Aeronautics and Space Administration (NASA)

U.S. Department of Energy (USDOE)

U.S. Department of Agriculture (USDA)

Monsanto Corporation

AGENDA

Wed., Dec. 2 — Arrivals, registration, open buffet

Thur., Dec. 3 — **Plenary Session 1 – Background** — **Chair – Norman Rosenberg, PNNL**

0800	Welcome, logistics, statement of the issue, purposes and products of the workshop
0845	Kyoto, Bonn and Buenos Aires: Where does it stand ? Sandra Brown, Winrock Fdn.
9:15	A Technology Strategy for climate change mitigation John Clarke, PNNL
	Plenary Session 2 – Science Needs and New Technology
9:45	White Paper (1) Presentation Blaine Metting, PNNL and Jeff Smith, USDA
10:30	Break
10:45	Commentaries on Paper 1 William Schlesinger, Duke Univ.; Cynthia Cambardella, USDA/ARS; Craig Cox, SWCS; Jeff Amthor, ORNL
	General discussion of Paper 1
12:30	Lunch
	Plenary Session 3 – Monitoring and Verification **Chair – Roger Dahlman, DOE**
1:30	White Paper (2) Presentation Mac Post, ORNL, Cesar Izaurralde, PNNL, L. Mann. ORNL and Norman Bliss, USGS
2:15	Commentaries on Paper 2 Con Campbell, Agriculture Canada; Shashi Verma, Univ. of Nebraska; Douglas Johnson, Clark Univ. and Herman Mayeux, USDA
	General Discussion of Paper 2
4:00	Break
	Plenary Session 4 – Desertification and Soil Carbon Sequestration **Chair –Steve Hildebrand, ORNL**
4:15	White Paper (3) Presentation Rattan Lal, Ohio State Univ., Julian Dumanski and Hassan Hassan, World Bank
5:00	Commentaries on Paper 3 Larry Wilding, Texas A&M; Pierre Crosson, Resources for the Future; Charles Hutchinson, Univ. of Arizona and NASA; Robert Jackson, Texas A&M Univ.
	General Discussion of Paper 3
6:30	Adjourn
7:30	Reception and Dinner

Friday, Dec. 4

0830 **Working Group Breakout Session #1**

Group 1 (Science) Coordinators: D. Shriner, USDA/FS; J. Reilly, MIT; W. McGill, U. Alberta; G. Gee, PNNL

Group 2 (Monitoring) CA Jones, TAMU; J. Jastrow, ANL; H. Bolton, PNNL; D. Zilberman, UCal-Berkeley

Group 3 (Desertification) A. Morris, Treasury Dept.; J. Kimball, NRCS; W. Pennell, PNNL; C. Pieri, World Bank; L. Tieszen, USGS

11:00 **Plenary Session 5 -**

Working Group Reports to Plenary Discussion **Chair—Peter Kuch, EPA**

12:30 Lunch

1:30 **Plenary Session 6 – Economics and Policy** **Chair –TBA**

White Paper (4) Presentation Gregg Marland, ORNL and B. McCarl, Texas A&M Univ.

2:15 Commentaries on Paper 4
Debbie Reed, Office of Senator Kerrey; John McClelland, National Corn Growers Assn.; Ron Sands, PNNL; John Bennett, Sask. Soil Conservation Society.

General Discussion of Paper 4

4:15 Break

4:30 **Working Group Breakout Session #2**
To discuss economics and policy implications for (1) New Science; (2) Monitoring and Verification; (3) Carbon Sequestration and Desertification

6:30 Adjourn

7:30 Cash bar, dinner and entertainment

Sat., Dec. 5 **Plenary Session 7 – Wrapping Up** **Chair —Richard Stuckey, CAST**

0800 The DOE Carbon Sequestration Roadmap Gary Jacobs, ORNL

0830 Perspectives –A forum Participants: P. Kuch, USEPA; L. Drinkwater, Rodale; W. Breed, USDOE; W. Lindwall, Agr. Canada; A. Donnelly, GEMco; A. Morris, Treasury Dept.; T. Peterson, White House; A. Wrona, Nat'l Cotton Council; TBA, TBA

10:30 Break

11:00 Working Group Reports to Plenary

11:45 General discussion of report and publication plans, follow-up activities and actions, final words for the good of the cause.

12:30 Adjourn

Appendix III

Participants in St. Michaels Workshop

Name	Affiliation	Working Group	WG No.
Alesii, Bruno	Monsanto	New Science	1
Amthor, Jeffrey	ORNL	New Science	1
Anderson, Margot	USDA	Monitoring and Verification	2
Bennett, John	Producer	Monitoring and Verification	2
Berc, Jeri	NRCS	Desertification	3
Bliss, Norman	EROS Data Center	Monitoring and Verification	2
Bloomfield, Janine	Environmental Defense Fund	Desertification	3
Bolton, Harvey	PNNL	Monitoring and Verification	2
Braatz, Barbara	ICF	Monitoring and Verification	2
Breed, William	DOE	Monitoring and Verification	2
Brown, Robert	PNNL	Monitoring and Verification	2
Brown, Sandra	Winrock International	Desertification	3
Buford, Marilyn	US Forest Service	New Science	1
Cambardella, Cynthia	USDA	New Science	1
Campbell, Con	Agric. & Agri-Food Canada	Monitoring and Verification	2
Clarke, John	PNNL	New Science	1
Cole, Vernon	Colorado St. Univ.	Desertification	3
Cox, Craig	SWCS	New Science	1
Crookshank, Steve	API	New Science	1
Crosson, Pierre	Resources for the Future	Desertification	3
Dahlman, Roger	DOE	New Science	1
Dick, Mike	Producer	Desertification	3
Donnelly, Aldyen	GEMCo	Monitoring and Verification	2
Drinkwater, Laurie	Rodale Inst.	New Science	1
Dumanski, Julian	World Bank	Desertification	3
Eisenberg, Jeff	The Nature Conservancy	New Science	1
Elwood, Jerry	DOE	Monitoring and Verification	2
Faeth, Paul	World Resources Inst.	Monitoring and Verification	2
Farrell, Michael	ORNL	Desertification	3
Frome, Michele	SWCS	Monitoring and Verification	2
Gassman, Philip	Iowa St. Univ.	Desertification	3
Gee, Glendon	PNNL	New Science	1

Name	Affiliation	Working Group	WG No.
Goss, Donald	Texas A&M Univ.	Desertification	3
Graham, Robin	ORNL	Desertification	3
Harden, John	Am. Soybean Assoc.	Monitoring and Verification	2
Harmon, Mark	Oregon St. Univ.	Monitoring and Verification	2
Hendrey, George	BNL	Monitoring and Verification	2
Hickman, John	John Deere Co.	Monitoring and Verification	2
Hildebrand, Steve	ORNL	Desertification	3
House, Robert	USDA	Monitoring and Verification	2
Hutchinson, Charles	NASA	Desertification	3
Izaurralde, Cesar	PNNL	Monitoring and Verification	2
Jackson, Robert	Univ. of Texas	Desertification	3
Jacobs, Gary	ORNL	New Science	1
Janetos, Anthony	NASA	Monitoring and Verification	2
Jastrow, Julie	ANL	Monitoring and Verification	2
Johnson, Douglas	Clark Univ.	Monitoring and Verification	2
Jones, Allan	Texas A&M Univ.	Monitoring and Verification	2
Kimble, John	NRCS	Desertification	3
Kuch, Peter	EPA	Desertification	3
Lal, Rattan	Ohio St. Univ.	Desertification	3
Lindwall, Wayne	Agric. & Agri-Food Canada	Monitoring and Verification	2
Loh, Hsue-Peng	DOE	New Science	1
Manale, Andrew	EPA	New Science	1
Marland, Gregg	ORNL	Desertification	3
Mayeux, Herman	USDA	Monitoring and Verification	2
McBeth, Daryn	Am. Soybean Assoc.	New Science	1
McCarl, Bruce	Texas A&M Univ.	New Science	1
McCarthy, John	ORNL	Monitoring and Verification	2
McClelland, John	Corn Growers Assoc.	New Science	1
McGill, William	Univ. of Alberta	New Science	1
Mercier, Stephanie	US Congress	Desertification	3
Metting, Blaine	PNNL	New Science	1
Morris, Adele	CEQ	Desertification	3
Olsson, Lennert	Lund Univ.	Desertification	3
Ordoñez, Carlos	Univ. of Houston	New Science	1
Pennell, William	PNNL	Desertification	3
Peterson, Tom	White House	Desertification	3
Pieri, Christian	World Bank	Desertification	3
Post, Wilfred	ORNL	Monitoring and Verification	2
Rayner, Steve	PNNL	Desertification	3
Reed, Debbie	US Senate	Monitoring and Verification	2
Reilly, John	MIT	New Science	1
Riordan, Courtney	CIA	Desertification	3
Rosenberg, Norman	PNNL	Desertification	3
Ruark, Greg	USDA	New Science	1
Sands, Ronald	PNNL	New Science	1
Schlesinger, William	Duke Univ.	New Science	1
Shriner, David	US Forest Service	New Science	1
Smith, Jeff	USDA	New Science	1
Smith, Megan	Am. Bioenergy Assoc.	New Science	1
Stokes, Bryce	US Forest Service	Monitoring and Verification	2
Stuckey, Richard	CAST	New Science	1

Name	Affiliation	Working Group	WG No.
Summers, Jeff	DOE	Monitoring and Verification	2
Tieszen, Larry	EROS Data Center	Desertification	3
Verma, Shashi	Univ. of Nebraska	Monitoring and Verification	2
von Grebmer, Klaus	Novartis Corp.	New Science	1
White, Bob	US Senate	New Science	1
Wilding, Larry	Texas A&M Univ.	Desertification	3
Williams, Jimmy	Texas A&M Univ.	Monitoring and Verification	2
Wirth, Tom	EPA	Desertification	3
Wright, Lynn	ORNL	New Science	1
Wrona, Anne	National Cotton Growers Assn.	Desertification	3
Yehle, Keith	US Senate	New Science	1
Zilberman, David	Univ. of California–Berkeley	Monitoring and Verification	2